THE
ART OF WAR

ILLUSTRATED EDITION

THE
ART OF WAR

ILLUSTRATED EDITION

孫子兵法
SUN TZǓ

FALL RIVER PRESS

New York

FALL RIVER PRESS

New York

An Imprint of Sterling Publishing
1166 Avenue of the Americas
New York, NY 10036

Cover design by Scott Russo
Book design by Gavin Motnyk

ISBN 978-1-4549-1186-9

Distributed in Canada by Sterling Publishing
c/o Canadian Manda Group, 664 Annette Street
Toronto, Ontario, Canada M6S 2C8

For information about custom editions, special sales, and premium and corporate
purchases, please contact Sterling Special Sales at 800-805-5489 or
specialsales@sterlingpublishing.com.

Manufactured in China

2   4   6   8   10   9   7   5   3

www.sterlingpublishing.com

CONTENTS

Introduction 8

Editor's Note 21

THE ART OF WAR

Preface 24

Introduction 27

LAYING PLANS 71

WAGING WAR 81

ATTACK BY STRATAGEM 91

TACTICAL DISPOSITIONS 105

ENERGY 113

WEAK POINTS AND STRONG 125

MANEUVERING 137

VARIATION IN TACTICS 153

THE ARMY ON THE MARCH 163

TERRAIN 181

THE NINE SITUATIONS 195

THE ATTACK BY FIRE 233

THE USE OF SPIES 245

APPENDICES

The Life and Times of Sun Tzǔ 259

Examples of Modern Translations 263

Based on the Book 266

Further Reading 269

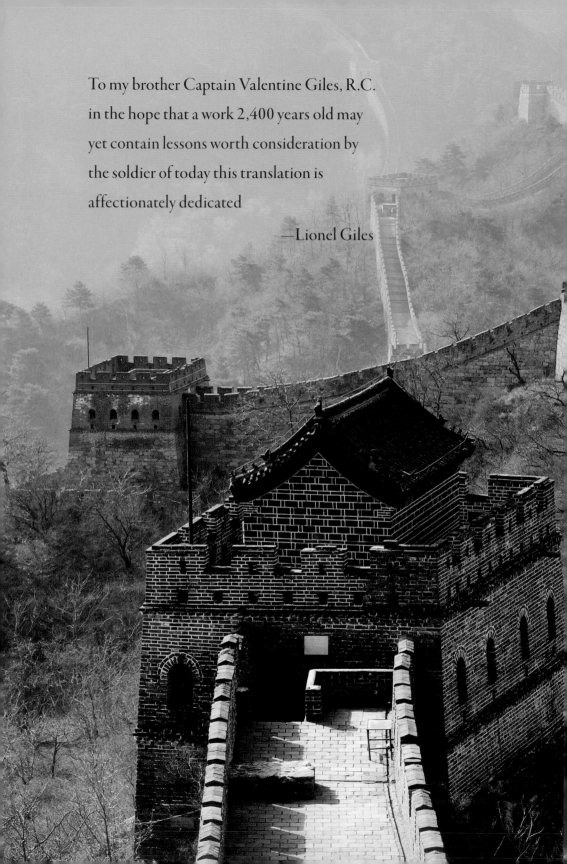

To my brother Captain Valentine Giles, R.C.
in the hope that a work 2,400 years old may
yet contain lessons worth consideration by
the soldier of today this translation is
affectionately dedicated

—Lionel Giles

INTRODUCTION

G iven the central role war has played in history, it may surprise that so few books on the conduct of war have achieved the status of classic. Even more astonishingly, one of the classic texts may well be the oldest known book on the subject. Widely believed to have been written 2,500 years ago, Sun Tzŭ's *The Art of War* continues to instruct generals across the world in the ways of winning wars. And its usefulness is not limited to military leaders: businessmen scour Sun Tzŭ for methods of managing their companies and defeating their competitors. Some editions even claim that reading the ancient Chinese master can help suitors win over the opposite sex and assist sports coaches in putting their teams on the road to victory.

How is it that this work has withstood the test of time so successfully and conquered the worlds of war, business, sports, and love? The quality of the message is, of course, one reason. But, as this introduction will explore, there are others. The uncertainties regarding authorship, the long gestation history of the text, the style of writing, and luck have all played equally important parts. In combination, these factors have allowed many generations of Chinese and Japanese, and much more recently, thanks to translation, the rest of the world, to construct "their" Sun Tzŭ.

According to tradition, Sun Tzŭ (Sunzi), or Master Sun—whose full name was Sun Wu—served as a general under Ho Lu (Helü), the king of Wu, who reigned from 514 to 496 BCE.[1] Details of his life are scanty. For all his later renown, Sun Tzŭ does not appear in any contemporary

孫子

---

1 I follow the older Wades-Giles transcription of Chinese, as this was used by the sinologist Lionel Giles, who assisted his father in establishing this romanization system and upon whose translation this edition of *The Art of War* is based. The current "pinyin" transcription is given in parentheses.

records. The first story about his life dates from some four hundred years later and takes the form of a morality tale. The great historian Ssŭ-ma Ch'ien (Sima Qian) (ca. 145 or 135 BCE–86 BCE) relates how King Ho Lu desired to test the author of "the thirteen chapters" on the art of war by having him train 180 ladies of the palace for war. Sun Tzŭ is said to have divided the group into two and appointed the king's favorite concubines to command each company. As the reader is meant to expect, the giggling women fail, not once but twice, to take the exercise seriously. Sun Tzŭ thereupon orders the summary execution of the company commanders. Ho Lu naturally objects, but is put in his place by Master Sun's stern comment that "the king is only fond of words, and cannot translate them into deeds." The concubines are executed and, in silence, the surviving women obey their general's every command. Convinced of his ability, Ho Lu appoints Sun Tzŭ his general and he "shared in the might of the King."[2] The story tells us little, if anything at all, about the personality of Sun Tzŭ. Instead it makes general points about the relative independence of military commanders from royal interference and the importance of iron discipline enforced through selective punishment. The lack of biographical information may well mean that Master Sun was a mythical figure whose reputation has never been in danger of being diminished by the knowledge of unpleasant facts about his life.

兵法

Lionel Giles, the noted sinologist responsible for the 1910 translation that forms the basis of this edition, nonetheless had little doubt that Sun Tzŭ was a real historical person and that his work was written between 505 and 473 BCE. Subsequent scholarship has been more doubtful. It is quite possible that the thirteen chapters ascribed to Sun Tzŭ are

---

2 For the full story, see the introduction to the translation by Lionel Giles on page 27.

actually a condensation of a number of writings that emerged over the centuries following Ho Lu's reign at the end of the Spring and Autumn Period (770–481 BCE) and to which the name of a single author was attached later. Scattered pieces of internal evidence, joined with a better recent understanding of developments in ancient Chinese warfare, indicate that the work's constituent parts probably date from a period in Chinese history known as the Warring States (481–221 BCE). Only a matter of decades ago, further supporting evidence emerged from a quite sensational archaeological discovery. In 1972, a tomb dating from the late second century BCE was opened up at Lin-i in Shantung Province in China. It contained a whole series of military manuscripts, including parts of each of the thirteen chapters that now make up Sun Tzŭ's *The Art of War*. The tomb did not conclusively settle the issue of authorship nor establish firmly the precise scope of Sun Tzŭ's writings. Master Sun is mentioned in some, but not all, fragments. Even more excitingly, and confusingly, his name also appears in fragments that were either unknown or not hitherto associated with him. The archaeological finds thus seem to indicate that writings credited to Sun Tzŭ formed over time and slowly coalesced around thirteen chapters.

兵
法

In fact, the text within the chapters remained remarkably open to change throughout its long history. This malleability was due to the ways in which the text was physically transmitted and intellectually handled. The ancient Chinese wrote single lines of vertical text on strips of bamboo or wood. These strips, usually some ten inches long and a third of an inch wide, were tied together, and rolled up in a bundle, they made a book. This kind of binding made them prone to breaking and scattering into single strips. In the tombs at Lin-i, archaeologists dug up a dirty, muddy jumble of 4,942 individual bamboo strips, which took literary scholars years to transcribe, organize, and publish. Interested readers can find a reconstruction of the Lin-i Sun Tzŭ in English by

Roger Ames—who, with D. C. Lau, recovered and translated another, hitherto completely lost work also entitled *The Art of War* by Sun Pin (Sun Bin), an author traditionally believed to be a descendant of Sun Tzŭ (see Further Reading on page 269). The work of modern scholars, like Ames and Lau, does not differ fundamentally from the practices of two millennia of Chinese scholars. Nowadays, we are perhaps more interested in establishing the original, authentic text and are able to disseminate it through authoritative printed editions with respectable publishers, but the editorial process continues to require decisions regarding the question of what constitutes "the" text. Editors of the past perhaps had more opportunities for introducing mistakes, as well as deliberate emendations and additions. For one thing, before the invention of printing, they needed to create each new copy by laboriously copying another by hand. Retying the strips of a book that had fallen apart could result in some parts being left out and forgotten, while here and there, new sections might be unobtrusively added. The vulnerability of the medium and its reproduction process thus allowed for a text over time to become removed from its original form.

孫
子

Chinese stylistic conventions added further opportunities for change. In line with the aphoristic nature of early Chinese writing, each strip tended to contain one or, at most, a couple of complete thoughts. This literary style tended not to value the systematic development of an argument across a number of strips. Since phrases and sections were also not numbered, trying to group thoughts under a general heading was no easy task for those who sought to put a broken book back together or transmit a corrupted copy. Even compilers and copyists who were trying to be faithful to the original faced a constant temptation to rearrange text, cut sections, and add interpolations to connect or explain passages that might seem unrelated or obscure. As was the case with the already established use of more perishable silk for writing, the invention of paper in

China in the second century CE did not change this. Perhaps because more text could now be seen on a page, it actually stimulated an editorial propensity for tinkering. It also made it much easier for comments by learned scholars to be added to the text. From this time onward, we see the emergence of a distinguished group of commentators, starting with one of China's greatest generals, Ts'ao Ts'ao (Cao Cao; 155–220 CE).

The later invention of book printing, by using wood blocks, marked an important next stage by allowing for a particular text version to become more fixed. It also meant that far more copies of a work could be produced, and so it would likely find wider dissemination. Thus, a key date for "the Sun Tzŭ," as the scholars in the field tend to say, was the publication of the first printed edition around the year 1200 CE during the emperor Ning-tsung's reign. This was based on a text that had been established by imperial edict more than a century earlier, when the Sun dynasty emperor Shen Tsung (reigned 1067–1085) had made the "Seven Military Classics" obligatory reading for those who wanted to obtain a commission in the army.[3] Printed editions also could easily accommodate the extensive commentaries which, as said, Sun Tzŭ had already been attracting since the second century CE. They could be interspersed with the short phrases of Sun Tzŭ's main text using a clearly recognizable, different size typeface. With time, however, copies of the Sung edition vanished, while other textual traditions survived and developed. By the late eighteenth century, the imperial government once again saw fit to try to establish order. Another official government edition was produced, this one by the antiquarian and classical scholar Sun Hsing-yen (1752–1818). This version became the standard text used by all modern English editions including Giles' until the discovery, in 1972, of the much earlier fragments that had been frozen in time in the tomb at Lin-i.

兵
法

---

3 These have been published in English as *The Seven Military Classics of Ancient China*, trans. and ed. Ralph D. Sawyer with Mei-chün Sawyer (Boulder, CO: Westview Press, 1993).

One can see the effects of this long, disorderly adaptive historical process in the modern translations of Sun Tzŭ. In this edition, Giles sometimes notes when a sentence or a numbered paragraph appears out of place. When one puts Giles' Sun Tzŭ next to other modern English editions, one will see that occasionally translators have decided to omit certain sections they judge overly obscure or out of place. A comparison also reveals that the text within the thirteen chapters is often divided and numbered differently. And, of course, passages are translated very differently (see the section with examples from various modern translations on page 263). An English reader can thus acquire an immediate sense of the tortuous transmission of the text and the range of possible interpretations by comparing various translations. Such a comparison also offers a glimpse of the great challenge modern Chinese speakers face when they attempt to decipher the classical Chinese of more than two thousand years—even the characters and concepts that can still be recognized have often shifted in meaning.

孫
子

Yet, far from being a weakness, the editorial engagement and flexible readings of the text have led to a constant process of regeneration that can be said to have kept the Sun Tzŭ alive for more than two millennia. Integral to this process was the Chinese practice of commenting on the classics. Like the variations in the text, the commentaries suggested different ways of understanding and interpreting the text and gave its meaning a flexibility that enabled it to adapt to changing times. The Sun Tzŭ was fortunate in its commentators. Ts'ao Ts'ao was joined over time by others, including the celebrated poet Tu Mu (Du Mu; 803–852 CE). By the time the "Seven Military Classics" were chosen and the first printed edition came out, a key group of eleven commentators was deemed to exist. The printed edition helped further elevate these eleven individuals to canonic status. Subsequent editors, copyists, and readers incorporated those commentaries that they believed to be relevant

and insightful, and sometimes added their own views. Giles continued in this tradition by carefully choosing particular commentary and also including his own voice. The respected sinologist John Minford notes in his recent translation of Sun Tzŭ that Giles "joined the ranks of the traditional eleven commentators. He is the twelfth man."[4]

Despite the fact that he wrote more than a century ago, Giles' extensive use of examples from Western military history, from the then-recent Boer War to the fifth-century BCE Peloponnesian War, gives a greater immediacy to Sun Tzŭ's ideas than some of the Chinese commentary is likely to achieve with Western readers. The examples also serve to reinforce the claim that Sun Tzŭ is relevant to the understanding and fighting of all war. Modern scholarly editorial practices, which strive to restore the authentic text in pristine condition, may regard this traditional approach as an unwarranted interference with the text, but the older practice has helped make Sun Tzŭ a living text and facilitated its survival across the centuries.

In this edition, the reader is first presented with the unadorned translation of Sun Tzŭ by Lionel Giles. This approach seems worthwhile in part because, as John Minford also writes, Giles' translation "has stood the test of time admirably."[5] It can bear the scrutiny of being read on its own, unencumbered by commentary. Thus, the reader can more easily respond to Sun Tzŭ's repeated injunction to ponder, and even meditate on, his sayings for him or herself, and provide, as it were, a personal commentary. Presenting the text in this fashion has the additional advantage of bringing into focus three important elements in the quality and presentational style of the ideas contained in Sun Tzŭ. These have been elements of major importance in assuring the text's

---

4 John Minford, introduction to *The Art of War*, by Sun-tzu (Sunzi), trans. and ed. J. Minford (New York: Penguin, 2003), pp. xxxviii–xxxix.
5 Minford, introduction, pp. xxxviii–xxxix.

longevity. First, there is the remarkable balance struck between generality and specificity, which avoids relating propositions to historically contingent practices in ways that have made most ancient Chinese (and Western) military writings completely outdated. Even when such long-vanished military machines like "swift" and "heavy" chariots are mentioned (II, 1), it is in a context that makes the general, timeless point that war is expensive and that protracted war ruins the state. Second, the whole of *The Art of War* has a strong tendency to point out what one should strive for, rather than explain how one should achieve one's aim. One of the most celebrated ideas in Sun Tzŭ is the proposition that, in war, "supreme excellence consists in breaking the enemy's resistance without fighting" (III, 2). This, the treatise goes on to say, is best achieved by "balking the enemy plans." But there is little on how exactly to do this. Since the methods of war have tended to change more than the issues around which war has revolved, omissions of this kind seem a wise authorial choice. The student of Sun Tzŭ is thus merely given specific pointers as to what to ponder, but the solution, or the way one chooses to tread, must be one's own. Third, the repeated use of metaphor and symbolism invites the reader to use his or her imagination and seek to apply propositions to a broader range of situations than the author may have had in mind. An example is the likening of the process of finding weakness in the enemy to the flowing of water (VI, 29–33). Yet, even in cases where a proposition clearly refers to a specific reality, the reader can choose to interpret it as symbolic and representative of other situations. Consider the following example: In chapter IX, a series of visual pointers are presented that were intended to assist the field commander in judging the enemy's disposition and the state of his army—for instance, "If birds gather on any spot, it is unoccupied. Clamor by night betokens nervousness." The observations are directly informed by the physical realities of the ancient battlefield.

兵
法

Nonetheless, one can easily read the phrase as signifying something deeper about how one must observe the military enemy or, for that matter, any antagonist.

For all *The Art of War*'s inherent qualities, Sun Tzŭ's worldwide reputation also benefited from a good dose of unpredictable luck. Neither the first French translation of Sun Tzŭ by Father Amiot in 1772, nor the English one by Giles in 1910, succeeded in making Sun Tzŭ well-known and fashionable in the West. His first moment of fame came only in the 1960s, when a US Marine Corps general who had served in China made a new translation. Samuel B. Griffith managed to get the most famous strategic thinker of the day, B. H. Liddell Hart, to write a foreword. Liddell Hart saw in Sun Tzŭ an exponent of his own strategic ideas of the "indirect approach," which emphasized circumventing the enemy's main sources of strength and defeating him in ways that avoided the long and bloody attritional forms of Clausewitzian warfare that had marked World Wars I and II. Griffith, himself, saw the value and relevance of Sun Tzŭ somewhat differently. The successful and, to many Westerners, surprising and shocking Communist takeover of power in China in 1949 and the subsequent rise of Maoist-inspired revolutionary warfare in the Third World required, in Griffith's opinion, a thorough understanding of the sources of inspiration of Chairman Mao's way of war. Mao's verdict that the Sun Tzŭ could be equated to "scientific truth" gave the work an obvious central importance in cracking the secret of revolutionary insurgencies.

The US Armed Forces, however, took some time before heeding Griffith's advice. It was only in the aftermath of the lost war in Vietnam and the resulting attempt to reshape military education that Sun Tzŭ and Clausewitz, the author of the only other real classic on war, became a central part of the curriculum at the various War Colleges. While Liddell Hart had seen Clausewitz and Sun Tzŭ as representatives of opposing approaches to war, today they are taught in complementary fashion which once again shows the malleability of Sun Tzŭ, as well as Clausewitz. Sun Tzŭ is valued for the emphasis on preparation of the battlefield and the advocacy of employing a broader range of means, in addition to brute force, such as deception, intelligence, and information warfare. These are all concepts that find their resonance in Sun Tzŭ. Perhaps more surprisingly, given the very different political environment in China two thousand years ago, even Sun Tzŭ's views on the relationship between government and military seem to have been prescient and appear correct to many. And with China finally emerging as a global superpower, Sun Tzŭ continues to hold out the promise of offering important insights into contemporary Chinese military thinking.

In the 1980s, Sun Tzŭ also benefited from the emergence of a new type of threat: the rise of Japan as an economic superpower. Sun Tzŭ had been known in Japan since the eighth century, at least according to tradition, and was a valued part of that nation's military heritage. Even though the Japanese understanding of Sun Tzŭ as reflected, for example, in the influential "Comparative Study of Sun Tzŭ and Clausewitz" by Lieutenant-General Muto Akira had perhaps not served them well in World War II (Akira was charged with involvement in the Nanjing massacre and executed as a war criminal), after the war, a suspicion arose in the West that the stunning Japanese economic successes might be due to Sun Tzŭ's influence. The result was a surge in business strategy books that were inspired by Sun Tzŭ. While the Japanese threat had waned by the 1990s, Sun Tzŭ had by then become a household name, and his work began to find adaptations in other markets.

Sun Tzŭ's fame appears assured and the uses of *The Art of War* manifold. The work's propositions regarding the wise way of conducting war and other conflictual situations which can variously be described as epigrammatic, apodictic, and laconic, or clever, obviously truthful, and short, continue to attract many. The possibility of winning without fighting remains an alluring idea, especially in casualty- and cost-averse democracies. Yet the common interpretation that Sun Tzŭ advocates achieving this through cynical calculation and an uncompromising manipulation of circumstances to one's own exclusive advantage should perhaps, in conclusion, be weighed against a possible alternate reading of the text. Sun Tzŭ does not advocate simple power politics and a winner-take-all approach to war. War, or more broadly, conflict is viewed as a never-ending process in Sun Tzŭ. "Winning" is never defined in the treatise or even identified as an attainable, conclusive state of affairs. Because no end state, no final peace, is ever reached, there will always be an opponent with whom to contend. The acceptance of an always existing enemy with whom one continually interacts violently can be taken to imply a need, or even a duty, to treat the opposition with respect and restraint. Paradoxically perhaps, the absence of a vision of peace in Sun Tzŭ means that there is no vision of unconditional surrender and extermination either.

Jan Willem Honig is senior lecturer in the Department of War Studies, King's College London. Between 2007 and 2011, he was professor of military strategy at the Swedish National Defence College. He writes extensively on strategy, and is the co-editor, with Andreas Herberg-Rothe and Daniel Moran, of Clausewitz: The State and War.

EDITOR'S NOTE

Lionel Giles (1875–1958) was a Victorian scholar who served as assistant curator at the British Museum and Keeper of the Department of Oriental Manuscripts and Printed Books. His many translations of Chinese texts include Sun Tzŭ's The Art of War *(1910),* The Analects of Confucius *(1910), and Lao Tzŭ's* Tao Te Ching *(1912). Giles' translation of* The Art of War *superseded other Western translations of the text between 1772 and 1905 that were largely discredited for their omissions and distortions. It became the standard English translation of Sun Tzŭ's treatise for nearly half a century.*

The *Art of War* text presented here is based on Giles's translation *Sun Tzu on the Art of War* (London: Luzac & Co., 1910). In Giles's selected commentary, elisions are indicated by ellipses (...) and paraphrases are bracketed.

兵
法

Spelling and punctuation have been Americanized throughout.

The "Preface" that follows, which appeared in Lionel Giles' original 1910 translation, discusses several earlier translations for which his was intended as a corrective. In his "Introduction," Giles acknowledges confusion over the identity of Sun Tzŭ caused by conflation of Sun Tzŭ with other historical personages in the work of Chinese historians, and by the attribution of Sun Tzŭ's teachings to texts published in the centuries after *The Art of War* appeared. Giles' meticulous scholarly study of references to Sun Tzŭ in historical writings not only confirmed the reality of Sun Tzŭ as an actual person (many writers considered him a figure of myth and folklore) but established *The Art of War* as a work of the 6th century BC. Historical figures and books mentioned in Giles' "Introduction" and the commentary he interpolated into his translation of Sun Tzŭ's treatise are described in the sections "The Commentators" and "Bibliography" towards the end of his "Introduction."

兵
法
THE
ART OF WAR

(with commentary)

PREFACE

The seventh volume of *Memoires concernant l'histoire, les sciences, les arts, les mœurs, les usages, &c., des Chinois* is devoted to *The Art of War*, and contains, amongst other treatises, "Les Treize Articles de Sun-tse," translated from the Chinese by a Jesuit Father, Joseph Amiot. Père Amiot appears to have enjoyed no small reputation as a sinologue in his day, and the field of his labors was certainly extensive. But his so-called translation of Sun Tzŭ, if placed side by side with the original, is seen at once to be little better than an imposture. It contains a great deal that Sun Tzŭ did not write, and very little indeed of what he did. . . .

Throughout the nineteenth century, which saw a wonderful development in the study of Chinese literature, no translator ventured to tackle Sun Tzŭ, although his work was known to be highly valued in China as by far the oldest and best compendium of military science. It was not until the year 1905 that the first English translation, by Captain E. F. Calthrop, R.F.A., appeared at Tokyo under the title "Sonshi" (the Japanese form of Sun Tzŭ). Unfortunately, it was evident that the translator's knowledge of Chinese was far too scanty to fit him to grapple with the manifold difficulties of Sun Tzŭ. He himself plainly acknowledges that without the aid of two Japanese gentlemen "the accompanying translation would have been impossible." We can only wonder, then, that with their help it should have been so excessively bad. It is not merely a question of downright blunders, from which none can hope to be wholly exempt. Omissions were frequent; hard passages were wilfully distorted or slurred over. Such offences are less pardonable. They would not be tolerated in any edition of a Greek or Latin classic, and a similar standard of honesty ought to be insisted upon in translations from Chinese.

From blemishes of this nature, at least, I believe that the present translation is free. It was not undertaken out of any inflated estimate of my own powers; but I could not help feeling that Sun Tzŭ deserved a better fate than had befallen him, and I knew that, at any rate, I could hardly fail to improve on the work of my predecessors. Towards the end of 1908, a new and revised edition of Captain Calthrop's translation was published in London, this time however, without any allusion to his Japanese collaborators. My first three chapters were then already in the printer's hands, so that the criticisms of Capt. Calthrop therein contained must be understood as referring to his earlier edition. In the subsequent chapters I have of course transferred my attention to the second edition. This is on the whole an improvement on the other, though there still remains much that cannot pass muster. Some of the grosser blunders have been rectified and lacunae filled up, but on the other hand a certain number of new mistakes appear. The very first sentence of the introduction is startlingly inaccurate; and later on, while mention is made of "an army of Japanese commentators" on Sun Tzŭ (who are these, by the way?), not a word is vouchsafed about the Chinese commentators, who nevertheless, I venture to assert, form a much more numerous and infinitely more important "army."

A few special features of the present volume may now be noticed. In the first place, the text has been cut up into numbered paragraphs, both in order to facilitate cross-reference and for the convenience of students generally. The division follows broadly that of Sun Hsing-yen's edition; but I have sometimes found it desirable to join two or more of his paragraphs into one. In quoting from other works, Chinese writers seldom give more than the bare title by way of reference, and the task of research is apt to be seriously hampered in consequence. With a view to obviating this difficulty so far as Sun Tzŭ is concerned, I have also appended a complete concordance of Chinese characters, following in this the admirable example of Legge, though an alphabetical arrangement has been preferred to the distribution under radicals which he adopted. Another feature borrowed from the Chinese classics is the printing of text, translation and notes on the same page; the notes, however, are inserted, according to the Chinese method, immediately after the passages to which they refer. From the mass of native commentary my aim has been to extract the cream only, . . . Though constituting in itself an important branch of Chinese literature, very little commentary of this kind has hitherto been made directly accessible by translation.

INTRODUCTION

Sun Wu and His Book
Ssŭ-ma Ch'ien gives the following biography of Sun Tzŭ:

Sun Tzŭ Wu was a native of the Ch'i State. His *Art of War* brought him to the notice of Ho Lü, King of Wu. Ho Lü said to him: "I have carefully perused your 13 chapters. May I submit your theory of managing soldiers to a slight test?"

Sun Tzŭ replied: "You may."

Ho Lü asked: "May the test be applied to women?"

The answer was again in the affirmative, so arrangements were made to bring 180 ladies out of the Palace. Sun Tzŭ divided them into two companies, and placed one of the King's favorite concubines at the head of each. He then bade them all take spears in their hands, and addressed them thus: "I presume you know the difference between front and back, right hand and left hand?"

The girls replied: "Yes."

Sun Tzŭ went on: "When I say 'Eyes front,' you must look straight ahead. When I say 'Left turn,' you must face towards your left hand. When I say 'Right turn,' you must face towards your right hand. When I say 'About turn,' you must face right round towards your back."

Again the girls assented. The words of command having been thus explained, he set up the halberds and battle-axes in order to begin the drill. Then, to

兵
法

the sound of drums, he gave the order "Right turn." But the girls only burst out laughing. Sun Tzŭ said: "If words of command are not clear and distinct, if orders are not thoroughly understood, then the general is to blame."

So he started drilling them again, and this time gave the order "Left turn," whereupon the girls once more burst into fits of laughter. Sun Tzŭ: "If words of command are not clear and distinct, if orders are not thoroughly understood, the general is to blame. But if his orders *are* clear, and the soldiers nevertheless disobey, then it is the fault of their officers."

So saying, he ordered the leaders of the two companies to be beheaded. Now the king of Wu was watching the scene from the top of a raised pavilion; and when he saw that his favorite concubines were about to be executed, he was greatly alarmed and hurriedly sent down the following message: "We are now quite satisfied as to our general's ability to handle troops. If we are bereft of these two concubines, our meat and drink will lose their savor. It is our wish that they shall not be beheaded."

Sun Tzŭ replied: "Having once received His Majesty's commission to be the general of his forces, there are certain commands of His Majesty which, acting in that capacity, I am unable to accept."

Accordingly, he had the two leaders beheaded, and straightway installed the pair next in order as leaders in their place. When this had been done, the drum was sounded for the drill once more; and the girls went through all the evolutions, turning to the right or to the left, marching ahead or wheeling back, kneeling or standing, with perfect accuracy and precision, not venturing to utter a sound. Then Sun Tzŭ sent a messenger to the King saying: "Your soldiers, Sire, are now properly drilled and disciplined, and ready for your majesty's inspection. They can be put to any use that their sovereign may desire; bid them go through fire and water, and they will not disobey."

But the King replied: "Let our general cease drilling and return to camp. As for us, we have no wish to come down and inspect the troops."

Thereupon Sun Tzŭ said: "The King is only fond of words, and cannot translate them into deeds."

After that, Ho Lü saw that Sun Tzŭ was one who knew how to handle an army, and finally appointed him general. In the west, he defeated the Ch'u State and forced his way into Ying, the capital; to the north he put fear into the States of Ch'i and Chin, and spread his fame abroad amongst the feudal princes. And Sun Tzŭ shared in the might of the King.

兵
法

About Sun Tzŭ himself this is all that Ssŭ-ma Ch'ien has to tell us in this chapter. But he proceeds to give a biography of his descendant, Sun Pin, born about a hundred years after his famous ancestor's death, and also the outstanding military genius of his time. The historian speaks of him too as Sun Tzŭ, and in his preface we read: "Sun Tzŭ had his feet cut off and yet continued to discuss the art of war." It seems likely, then, that "Pin" was a nickname bestowed on him after his mutilation, unless the story was invented in order to account for the name. The crowning incident of his career, the crushing defeat of his treacherous rival P'ang Chüan, will be found briefly related in [ch. V, sect. 19 of *The Art of War*]. To return to the elder Sun Tzŭ, he is mentioned in two other passages of the *Shih Chi*:

In the third year of his reign [512 BC] Ho Lü, king of Wu, took the field with Tzŭ-hsü [i.e., Wu Yüan] and Po P'ei, and attacked Ch'u. He captured the town of Shu and slew the two prince's sons who had formerly been generals of Wu. He was then meditating a descent on Ying [the capital]; but the general Sun Wu said: "The army is exhausted. It is not yet possible. We must wait." . . . [After further successful fighting,] "in the ninth year [506 BC], King Ho Lü addressed Wu Tzŭ-hsü and Sun Wu, saying: "Formerly, you declared that it was not yet possible for us to enter Ying. Is the time ripe now?" The two men replied: "Ch'u's general Tzŭ-ch'ang, is grasping and covetous, and the princes of T'ang and Ts'ai both have a grudge against him. If Your Majesty has resolved to make a grand attack, you must win over T'ang and Ts'ai, and then you may succeed." Ho Lü followed this advice [and beat Ch'u in five pitched battles and marched into Ying].

孫
子

This is the latest date at which anything is recorded of Sun Wu [another name by which Sun Tzŭ is known]. He does not appear to have survived his patron, who died from the effects of a wound in 496.

In another chapter there occurs this passage:

> From this time onward, a number of famous soldiers arose, one after the other: Kao-fan, who was employed by the Chin State; Wang-tzŭ, in the service of Ch'i; and Sun Wu, in the service of Wu. These men developed and threw light upon the principles of war.

It is obvious enough that Ssŭ-ma Ch'ien at least had no doubt about the reality of Sun Wu as an historical personage; and with one exception, to be noticed presently, he is by far the most important authority on the period in question. It will not be necessary, therefore, to say much of such a work as the *Wu Yüeh Ch'un Ch'iu*, which is supposed to have been written by Chao Yeh of the 1st century AD. The attribution is somewhat doubtful; but even if it were otherwise, his account would be of little value, based as it is on the *Shih Chi* and expanded with romantic details. The story of Sun Tzŭ will be found, for what it is worth, in chapter 2. The only new points in it worth noting are: (1) Sun Tzŭ was first recommended to Ho Lü by Wu Tzŭ-hsü. (2) He is called a native of Wu. (3) He had previously lived a retired life, and his contemporaries were unaware of his ability.

The following passage occurs in Huai-nan Tzŭ: "When sovereign and ministers show perversity of mind, it is impossible even for a Sun Tzŭ to encounter the foe." Assuming that this work is genuine (and hitherto no doubt has been cast upon it), we have here the earliest direct reference for Sun Tzŭ, for Huai-nan Tzŭ died in 122 BC, many years before the *Shih Chi* was given to the world.

Liu Hsiang (80–89 BC) says: "The reason why Sun Tzŭ at the head of 30,000 men beat Ch'u with 200,000 is that the latter were undisciplined."

Têng Ming-shih . . . informs us that the surname [Sun] was bestowed on Sun Wu's grandfather by Duke Ching of Ch'i [547–490 BC]. Sun Wu's father Sun P'ing, rose to be a Minister of State in Ch'i, and Sun Wu himself, whose style was Ch'ang-ch'ing, fled to Wu on account of the rebellion which was being fomented by the kindred of T'ien Pao. He had three sons, of whom the second, named Ming, was the father of Sun Pin. According to this account then, Pin was the grandson of Wu, which, considering that Sun Pin's victory over Wei was gained in 341 BC, may be dismissed as chronologically impossible. Whence these data were obtained by Têng Ming-shih I do not know, but of course no reliance whatever can be placed in them.

An interesting document which has survived from the close of the Han period is the short preface written by the Great Ts'ao Ts'ao, or Wei Wu Ti, for his edition of Sun Tzŭ. I shall give it in full:

孫
子

> I have heard that the ancients used bows and arrows to their advantage. The *Lun Yü* says: "There must be a sufficiency of military strength." The *Shu Ching* mentions "the army" among the "eight objects of government." The *I Ching* says: "'army' indicates firmness and justice; the experienced leader will have good fortune." The *Shih Ching* says: "The King rose majestic in his wrath, and he marshaled his troops." The Yellow Emperor, T'ang the Completer and Wu Wang all used spears and battle-axes in order to succor their generation. The *Ssŭ-ma Fa* says: "If one man slay another of set purpose, he himself may rightfully be slain." He who relies solely on warlike measures shall be exterminated; he who relies solely on peaceful measures shall perish. Instances of this are Fu Ch'ai on the one hand and Yen Wang on the other. In military matters, the Sage's rule is normally to keep the peace, and to move his forces only when occasion requires. He will not use armed force unless driven to it by necessity.
>
> Many books have I read on the subject of war and fighting; but the work composed by Sun Wu is the profoundest of them all. [Sun Tzŭ was a native of the Ch'i state, his personal name was Wu. He wrote the *Art of War* in 13 chapters for Ho Lü, King of Wu. Its principles were tested on women, and he was subsequently made a general. He led an army westwards, crushed the Ch'u state and entered Ying the capital. In the north, he kept Ch'i and Chin in awe. A hundred years and more after his time, Sun Pin lived. He was a descendant of Wu.] In his treatment of

deliberation and planning, the importance of rapidity in taking the field, clearness of conception, and
depth of design, Sun Tzŭ stands beyond the reach of
carping criticism. My contemporaries, however, have
failed to grasp the full meaning of his instructions,
and while putting into practice the smaller details
in which his work abounds, they have overlooked its
essential purport. That is the motive which has led
me to outline a rough explanation of the whole.

One thing to be noticed in the above is the explicit statement that the
13 chapters were specially composed for King Ho Lü. This is supported
by the internal evidence of ch. I, sect. 15, in which it seems clear that
some ruler is addressed.

In the bibliographical section of the *Han Shu*, there is an entry which
has given rise to much discussion: "The works of Sun Tzŭ of Wu in
82 *p'ien* (or chapters), with diagrams in 9 *chüan*." It is evident that this
cannot be merely the 13 chapters known to Ssŭ-ma Ch'ien, or those we
possess today. Chang Shou-chieh . . . refers to an edition of Sun Tzŭ's
[*The Art of War*] of which the "13 chapters" formed the first *chüan*, add-
ing that there were two other *chüan* besides. This has brought forth a
theory, that the bulk of these 82 chapters consisted of other writings
of Sun Tzŭ—we should call them apocryphal—similar to the *Wên Ta*,
of which a specimen dealing with the Nine Situations is preserved in
the *T'ung Tien*, and another in Ho Shih's commentary. It is suggested
that before his interview with Ho Lu, Sun Tzŭ had only written the 13
chapters, but afterwards composed a sort of exegesis in the form of question and answer between himself and the King. Pi I-hsün, the author
of the *Sun Tzŭ Hsü Lu*, backs this up with a quotation from the *Wu
Yüeh Ch'un Ch'iu*: "The King of Wu summoned Sun Tzŭ, and asked
him questions about the art of war. Each time he set forth a chapter of

his work, the King could not find words enough to praise him." As he points out, if the whole work was expounded on the same scale as in the above-mentioned fragments, the total number of chapters could not fail to be considerable. Then the numerous other treatises attributed to Sun Tzŭ might be included. The fact that the *Han Chih* mentions no work of Sun Tzŭ except the 82 *p'ien*, whereas the Sui and T'ang bibliographies give the titles of others in addition to the "13 chapters," is good proof, Pi I-hsün thinks, that all of these were contained in the 82 *p'ien*. Without pinning our faith to the accuracy of details supplied by the *Wu Yüeh Ch'un Ch'iu*, or admitting the genuineness of any of the treatises cited by Pi I-hsün, we may see in this theory a probable solution of the mystery. Between Ssŭ-ma Ch'ien and Pan Ku there was plenty of time for a luxuriant crop of forgeries to have grown up under the magic name of Sun Tzŭ, and the 82 *p'ien* may very well represent a collected edition of these lumped together with the original work. It is also possible, though less likely, that some of them existed in the time of the earlier historian and were purposely ignored by him.

Tu Mu, after Ts'ao Kung the most important commentator on Sun Tzŭ, composed the preface to his edition about the middle of the ninth century. After a somewhat lengthy defense of the military art, he comes at last to Sun Tzŭ himself, and makes two startling assertions: "The writings of Sun Wu," he said, "originally comprise several hundred thousand words, but Ts'ao Ts'ao, the Emperor Wu Wei, pruned away all redundancies and wrote out the essence of the whole, so as to form a single book in 13 chapters." He goes on to remark that Ts'ao Ts'ao's commentary on Sun Wu leaves a certain proportion of difficulties unexplained. This, in Tu Mu's opinion, does not necessarily imply that he was unable to furnish a complete commentary. According to the *Wei Chih*, Ts'ao himself wrote a book on war in something over 100,000 words, known as the *Hsin Shu*. It appears to have been of such exceptional merit that he suspects Ts'ao to have used for it the surplus material he found in Sun Tzŭ. He concludes, however, by saying "the *Hsin Shu* is now lost, so that the truth cannot be known for certain."

Tu Mu's conjecture seems to be based on a passage which states: "Wei Wu Ti strung together Sun Wu's *Art of War*," which in turn may have resulted from a misunderstanding of the final words of Ts'ao King's preface. This, as Sun Hsing-yen points out, is only a modest way of saying that he made an explanatory paraphrase, or in other words, wrote a commentary on it. On the whole, this theory has met with very little acceptance. Thus, the *Ssu K'u Ch'uan Shu* says: "The mention of the 13 chapters in the Shih Chi shows that they were in existence before the *Han Chih*, and that latter accretions are not to be considered part of the original work. Tu Mu's assertion can certainly not be taken as proof."

There is every reason to suppose, then, that the 13 chapters existed in the time of Ssŭ-ma Ch'ien practically as we have them now. That the work was then well known he tells us in so many words. "Sun Tzŭ's 13 Chapters and Wu Ch'i's *Art of War* are the two books that people commonly refer to on the subject of military matters. Both of them are widely distributed, so I will not discuss them here." But as we go further back, serious difficulties begin to arise. The salient fact which has to be faced is that the *Tso Chuan*, the greatest contemporary record, makes no mention whatsoever of Sun Wu, either as a general or as a writer. It is natural, in view of this awkward circumstance, that many scholars should not only cast doubt on the story of Sun Wu as given in the *Shih Chi*, but even show themselves frankly skeptical as to the existence of the man at all. The most powerful presentment of this side of the case is to be found in the following disposition by Yeh Shui-hsin (1151–1223):

> It is stated in Ssŭ-ma Ch'ien's history that Sun
> Wu was a native of the Ch'i State, and employed by

兵
法

Wu; and that in the reign of Ho Lü he crushed Ch'u, entered Ying, and was a great general. But in Tso's Commentary no Sun Wu appears at all. It is true that Tso's Commentary need not contain absolutely everything that other histories contain. But Tso has not omitted to mention vulgar plebeians and hireling ruffians such as Ying K'ao-shu, Ts'ao Kuei, Chu Chih-wu and Chüan She-chu. In the case of Sun Wu, whose fame and achievements were so brilliant, the omission is much more glaring. Again, details are given, in their due order, about his contemporaries Wu Yüan and the Minister P'ei. Is it credible that Sun Wu alone should have been passed over?

In point of literary style, Sun Tzŭ's work belongs to the same school as *Kuan Tzŭ*, *Liu T'ao*, and the *Yüeh Yü* and may have been the production of some private scholar living towards the end of the "Spring and Autumn" or the beginning of the "Warring States" period. The story that his precepts were actually applied by the Wu State, is merely the outcome of big talk on the part of his followers.

From the flourishing period of the Chou dynasty down to the time of the "Spring and Autumn," all military commanders were statesmen as well, and the class of professional generals, for conducting external campaigns, did not then exist. It was not until the period of the "Six States" that this custom changed. Now although Wu was an uncivilized State, is it conceivable that Tso should have left unrecorded the fact that Sun Wu was a great general and yet held no civil office? What we are told, therefore, about Jang-chü and Sun Wu, is not authentic matter, but the reckless fabrication of theorizing pundits. The story of Ho Lu's experiment on the women, in particular, is utterly preposterous and incredible.

Yeh Shui-hsin represents Ssŭ-ma Ch'ien as having said that Sun Wu crushed Ch'u and entered Ying. This is not quite correct. No doubt the impression left on the reader's mind is that he at least shared in these exploits. . . . The fact may or may not be significant; but it is nowhere explicitly stated in the *Shih Chi* either that Sun Tzŭ was general on the occasion of the taking of Ying, or that he even went there at all. Moreover, as we know that Wu Yüan and Po P'ei both took part in the expedition, and also that its success was largely due to the dash and enterprise of Fu Kai, Ho Lu's younger brother, it is not easy to see how yet another general could have played a very prominent part in the same campaign.

Ch'ên Chên-sun of the Sung dynasty has the note:

> Military writers look upon Sun Wu as the father of their art. But the fact that he does not appear in the Tso Chüan, although he is said to have served under Ho Lü King of Wu, makes it uncertain what period he really belonged to.

He also says:

> The works of Sun Wu and Wu Ch'i may be of genuine antiquity.

It is noticeable that both Yeh Shui-hsin and Ch'ên Chên-sun, while rejecting the personality of Sun Wu as he figures in Ssŭ-ma Ch'ien's history, are inclined to accept the date traditionally assigned to the work which passes under his name. The author of the *Hsü Lu* fails to appreciate this distinction, and consequently his bitter attack on Ch'ên Chên-sun really misses its mark. He makes one of two points, however, which certainly tell in favor of the high antiquity of our "13 chapters." "Sun Tzŭ," he says, "must have lived in the age of Ching

兵
法

Wang [519–476], because he is frequently plagiarized in subsequent works of the Chou, Ch'in and Han dynasties." The two most shameless offenders in this respect are Wu Ch'i and Huai-nan Tzǔ, both of them important historical personages in their day. The former lived only a century after the alleged date of Sun Tzǔ, and his death is known to have taken place in 381 BC. It was to him, according to Liu Hsiang, that Tsêng Shên delivered the *Tso Chuan*, which had been entrusted to him by its author. Now the fact that quotations from the *Art of War*, acknowledged or otherwise, are to be found in so many authors of different epochs, establishes a very strong anterior to them all—in other words, that Sun Tzǔ's treatise was already in existence towards the end of the 5th century BC. Further proof of Sun Tzǔ's antiquity is furnished by the archaic or wholly obsolete meanings attaching to a number of the words he uses. A list of these, which might perhaps be extended, is given in the *Hsü Lu*; and though some of the interpretations are doubtful, the main argument is hardly affected thereby. Again, it must not be forgotten that Yeh Shui-hsin, a scholar and critic of the first rank, deliberately pronounces the style of the 13 chapters to belong to the early part of the fifth century. Seeing that he is actually engaged in an attempt to disprove the existence of Sun Wu himself, we may be sure that he would not have hesitated to assign the work to a later date had he not honestly believed the contrary. And it is precisely on such a point that the judgment of an educated Chinaman will carry most weight. Other internal evidence is not far to seek. Thus in ch. XIII, sect. 1, there is an unmistakable allusion to the ancient system of land-tenure which had already passed away by the time of Mencius, who was anxious to see it revived in a modified form. The only warfare Sun Tzǔ knows is that carried on between the various feudal princes, in which armored chariots play a large part. Their use seems to have entirely died out before the end of the Chou dynasty. He speaks as a man of Wu, a state which ceased to exist as early as 473 BC. On this I shall touch presently.

But once refer the work to the 5th century or earlier, and the chances of its being other than a *bonâ fide* production are sensibly diminished. The great age of forgeries did not come until long after. That it should have been forged in the period immediately following 473 is particularly unlikely, for no one, as a rule, hastens to identify himself with a lost cause. As for Yeh Shui-hsin's theory, that the author was a literary recluse, that seems to me quite untenable. If one thing is more apparent than another after reading the maxims of Sun Tzŭ, it is that their essence has been distilled from a large store of personal observation and experience. They reflect the mind not only of a born strategist, gifted with a rare faculty of generalization, but also of a practical soldier closely acquainted with the military conditions of his time. To say nothing of the fact that these sayings have been accepted and endorsed by all the greatest captains of Chinese history, they offer a combination of freshness and sincerity, acuteness and common sense, which quite excludes the idea that they were artificially concocted in the study. If we admit, then, that the 13 chapters were the genuine production of a military man living towards the end of the "*Ch'un Ch'iu*" period, are we not bound, in spite of the silence of the *Tso Chuan*, to accept Ssŭ-ma Ch'ien's account in its entirety? In view of his high repute as a sober historian, must we not hesitate to assume that the records he drew upon for Sun Wu's biography were false and untrustworthy? The answer, I fear, must be in the negative. There is still one grave, if not fatal, objection to the chronology involved in the story as told in the *Shih Chi*, which, so far as I am aware, nobody has yet pointed out. There are two passages in Sun Tzŭ in which he alludes to contemporary affairs. The first is in ch. VI, sect. 21:

兵
法

Though according to my estimate the soldiers of Yüeh exceed our own in number, that shall advantage them nothing in the matter of victory. I say then that victory can be achieved.

The other is in ch. XI, sect. 30:

Asked if an army can be made to imitate the shuai-jan, I should answer, Yes. For the men of Wu and the men of Yüeh are enemies; yet if they are crossing a river in the same boat and are caught by a storm, they will come to each other's assistance just as the left hand helps the right.

These two paragraphs are extremely valuable as evidence of the date of composition. They assign the work to the period of the struggle between Wu and Yüeh. So much has been observed by Pi I-hsün. But what has hitherto escaped notice is that they also seriously impair the credibility of Ssŭ-ma Ch'ien's narrative. As we have seen above, the first positive date given in connection with Sun Wu is 512 BC. He is then spoken of as a general, acting as confidential adviser to Ho Lü, so that his alleged introduction to that monarch had already taken place, and of course the 13 chapters must have been written earlier still. But at that time, and for several years after, down to the capture of Ying in 506, Ch'u and not Yüeh, was the great hereditary enemy of Wu. The two states, Ch'u and Wu, had been constantly at war for over half a century, whereas the first war between Wu and Yüeh was waged only in 510, and even then was no more than a short interlude sandwiched in the midst of the fierce struggle with Ch'u. Now Ch'u is not mentioned in the 13 chapters at all. The natural inference is that they were written at a time when Yüeh had become the prime antagonist of Wu, that is, after Ch'u had suffered the great humiliation of 506. At this point, a table of dates may be found useful.

孫
子

BCE

514   Accession of Ho Lü.

512   Ho Lü attacks Ch'u, but is dissuaded from entering Ying, the capital. *Shi Chi* mentions Sun Wu as general.

511   Another attack on Ch'u.

510   Wu makes a successful attack on Yüeh. This is the first war between the two states.

509 or 508   Ch'u invades Wu, but is signally defeated at Yü-chang.

506   Ho Lü attacks Ch'u with the aid of T'ang and Ts'ai. Decisive battle of Po-chü, and capture of Ying. Last mention of Sun Wu in *Shih Chi*.

505   Yüeh makes a raid on Wu in the absence of its army. Wu is beaten by Ch'in and evacuates Ying.

504   Ho Lü sends Fu Ch'ai to attack Ch'u.

497   Kou Chien becomes King of Yüeh.

496   Wu attacks Yüeh, but is defeated by Kou Chien at Tsui-li. Ho Lü is killed.

494   Fu Ch'ai defeats Kou Chien in the great battle of Fu-chiao, and enters the capital of Yüeh.

485–484   Kou Chien renders homage to Wu. Death of Wu Tzŭ-hsü.

482   Kou Chien invades Wu in the absence of Fu Ch'ai.

478–476   Further attacks by Yüeh on Wu.

475   Kou Chien lays siege to the capital of Wu.

473   Final defeat and extinction of Wu.

兵法

The sentence quoted above from ch. VI, sect. 21 hardly strikes me as one that could have been written in the full flush of victory. It seems rather to imply that, for the moment at least, the tide had turned against Wu, and that she was getting the worst of the struggle. Hence we may conclude that our treatise was not in existence in 505, before which date Yüeh does not appear to have scored any notable success against Wu. Ho Lü died in 496, so that if the book was written for him, it must have been during the period 505–496, when there was a lull in the hostilities, Wu having presumably been exhausted by its supreme effort against Ch'u. On the other hand, if we choose to disregard the tradition connecting Sun Wu's name with Ho Lü, it might equally well have seen the light between 496 and 494, or possibly in the period 482–473, when Yüeh was once again becoming a very serious menace. We may feel fairly certain that the author, whoever he may have been, was not a man of any great eminence in his own day. On this point the negative testimony of the *Tso Tchuan* far outweighs any shred of authority still attaching to the *Shih Chi*, if once its other facts are discredited. Sun Hsing-yen, however, makes a feeble attempt to explain the omission of his name from the great commentary. It was Wu Tzŭ-hsü, he says, who got all the credit of Sun Wu's exploits, because the latter (being an alien) was not rewarded with an office in the State.

How then did the Sun Tzŭ legend originate? It may be that the growing celebrity of the book imparted by degrees a kind of factitious renown to its author. It was felt to be only right and proper that one so well versed in the science of war should have solid achievements to his credit as well. Now the capture of Ying was undoubtedly the greatest feat of arms in Ho Lü's reign; it made a deep and lasting impression on all the surrounding states, and raised Wu to the short-lived zenith of her power. Hence, what more natural, as time went on, than that the acknowledged master of strategy, Sun Wu, should be popularly

兵
法

identified with that campaign, at first perhaps only in the sense that his brain conceived and planned it; afterwards, that it was actually carried out by him in conjunction with Wu Yüan, Po P'ei and Fu Kai?

It is obvious that any attempt to reconstruct even the outline of Sun Tzŭ's life must be based almost wholly on conjecture. With this necessary proviso, I should say that he probably entered the service of Wu about the time of Ho Lü's accession, and gathered experience, though only in the capacity of a subordinate officer, during the intense military activity which marked the first half of the prince's reign. If he rose to be a general at all, he certainly was never on an equal footing with the three above mentioned. He was doubtless present at the investment and occupation of Ying, and witnessed Wu's sudden collapse in the following year. Yüeh's attack at this critical juncture, when her rival was embarrassed on every side, seems to have convinced him that this upstart kingdom was the great enemy against whom every effort would henceforth have to be directed. Sun Wu was thus a well-seasoned warrior when he sat down to write his famous book, which according to my reckoning must have appeared towards the end, rather than the beginning of Ho Lü's reign. The story of the women may possibly have grown out of some real incident occurring about the same time. As we hear no more of Sun Wu after this from any source, he is hardly likely to have survived his patron or to have taken part in the death-struggle with Yüeh, which began with the disaster at Tsui-li.

If these inferences are approximately correct, there is a certain irony in the fate which decreed that China's most illustrious man of peace should be contemporary with her greatest writer on war.

The Text of Sun Tzŭ

I have found it difficult to glean much about the history of Sun Tzŭ's text. The quotations that occur in early authors go to show that the "13 chapters" of which Ssŭ-ma Ch'ien speaks were essentially the same as those now extant. We have his word for it that they were widely circulated in his day, and can only regret that he refrained from discussing them on that account. Sun Hsing-yen says in his preface:

> During the Ch'in and Han dynasties Sun Tzŭ's
> *Art of War* was in general use amongst military com-
> manders, but they seem to have treated it as a work of
> mysterious import, and were unwilling to expound it
> for the benefit of posterity. Thus it came about that
> Wei Wu was the first to write a commentary on it.

As we have already seen, there is no reasonable ground to suppose that Ts'ao Kung tampered with the text. But the text itself is often so obscure, and the number of editions which appeared from that time onward so great, especially during the T'ang and Sung dynasties, that it would be surprising if numerous corruptions had not managed to creep in. Towards the middle of the Sung period, by which time all the chief commentaries on Sun Tzŭ were in existence, a certain Chi T'ien-pao published a work in 15 *chüan* entitled "Sun Tzŭ with the collected commentaries of ten writers." There was another text, with variant readings put forward by Chu Fu of Ta-hsing, which also had supporters among the scholars of that period; but in the Ming editions, Sun Hsing-yen tells us, these readings were for some reason or other no longer put into circulation. Thus, until the end of the 18th century, the text in sole possession of the field was one derived from Chi T'ien-pao's edition, although no actual copy of that important work was known to have survived. That, therefore, is the text of Sun

Tzŭ which appears in the War section of the great Imperial encyclopedia printed in 1726, the *Ku Chin T'u Shu Chi Ch'êng*. Another copy at my disposal of what is practically the same text, with slight variations, is that contained in the "Eleven philosophers of the Chou and Ch'in dynasties" (1758). And the Chinese printed in Capt. Calthrop's first edition is evidently a similar version which has filtered through Japanese channels.

So things remained until Sun Hsing-yen (1752–1818), a distinguished antiquarian and classical scholar, who claimed to be an actual descendant of Sun Wu, accidentally discovered a copy of Chi T'ien-pao's long-lost work, when on a visit to the library of the Hua-yin temple. Appended to it was the *I Shuo* of Chêng Yu-hsien, mentioned in the *T'ung Chih*, and also believed to have perished. This is what Sun Hsing-yen designates as the "original edition (or text)"—a rather misleading name, for it cannot by any means claim to set before us the text of Sun Tzŭ in its pristine purity. Chi T'ien-pao was a careless compiler, and appears to have been content to reproduce the somewhat debased version current in his day, without troubling to collate it with the earliest editions then available. Fortunately, two versions of Sun Tzŭ, even older than the newly discovered work, were still extant, one buried in the *T'ung Tien*, Tu Yu's great treatise on the Constitution, the other similarly enshrined in the *T'ai P'ing Yü Lan* encyclopedia. In both the complete text is to be found, though split up into fragments, intermixed with other matter, and scattered piecemeal over a number of different sections. Considering that the *Yü Lan* takes us back to the year 983, and the *T'ung Tien* about 200 years further still, to the middle of the T'ang dynasty, the value of these early transcripts of Sun Tzŭ can hardly be overestimated. Yet the idea of utilizing them does not seem to have occurred to anyone until Sun Hsing-yen, acting under Government instructions, undertook a thorough recension of the text. This is his own account:

孫
子

48

Because of the numerous mistakes in the text of Sun Tzŭ which his editors had handed down, the Government ordered that the ancient edition [of Chi T'ien-pao] should be used, and that the text should be revised and corrected throughout. It happened that Wu Nien-hu, the Governor Pi Kua, and Hsi, a graduate of the second degree, had all devoted themselves to this study, probably surpassing me therein. Accordingly, I have had the whole work cut on blocks as a textbook for military men.

The three individuals here referred to had evidently been occupied on the text of Sun Tzŭ prior to Sun Hsing-yen's commission, but we are left in doubt as to the work they really accomplished. At any rate, the new edition, when ultimately produced, appeared in the names of Sun Hsing-yen and only one co-editor Wu Jên-chi. They took the "original edition" as their basis, and by careful comparison with older versions, as well as the extant commentaries and other sources of information such as the *I Shuo*, succeeded in restoring a very large number of doubtful passages, and turned out, on the whole, what must be accepted as the closest approximation we are ever likely to get to Sun Tzŭ's original work. This is what will hereafter be denominated the "standard text."

The copy which I have used belongs to a reissue dated 1877. It is in 6 *pên*, forming part of a well-printed set of 23 early philosophical works in 83 *pên*. It opens with a preface by Sun Hsing-yen (largely quoted in this introduction), vindicating the traditional view of Sun Tzŭ's life and performances, and summing up in remarkably concise fashion the evidence in its favor. This is followed by Ts'ao Kung's preface to his edition, and the biography of Sun Tzŭ from the *Shih Chi*, both translated above. Then come, firstly, Chêng Yu-hsien's *I Shuo*, with author's preface, and

兵
法

next, a short miscellany of historical and bibliographical information entitled *Sun Tzŭ Hsü Lu*, compiled by Pi I-hsün. As regards the body of the work, each separate sentence is followed by a note on the text, if required, and then by the various commentaries appertaining to it, arranged in chronological order. These we shall now proceed to discuss briefly, one by one.

THE COMMENTATORS

Sun Tzŭ can boast an exceptionally long distinguished roll of commentators, which would do honor to any classic. Ou-yang Hsiu remarks on this fact, though he wrote before the tale was complete, and rather ingeniously explains it by saying that the artifices of war, being inexhaustible, must therefore be susceptible of treatment in a great variety of ways.

Ts'ao Ts'ao or *Ts'ao Kung*, afterwards known as *Wei Wu Ti* (AD 155–220). There is hardly any room for doubt that the earliest commentary on Sun Tzŭ actually came from the pen of this extraordinary man, whose biography in the *San Kuo Chih* reads like a romance. One of the greatest military geniuses that the world has seen, and Napoleonic in the scale of his operations, he was especially famed for the marvelous rapidity of his marches, which has found expression in the line "Talk of Ts'ao Ts'ao, and Ts'ao Ts'ao will appear." Ou-yang Hsiu says of him that he was a great captain who "measured his strength against Tung Cho, Lü Pu and the two Yüan, father and son, and vanquished them all; whereupon he divided the Empire of Han with Wu and Shu, and made himself king. It is recorded that whenever a council of war was held by Wei on the eve of a far-reaching campaign, he had all his calculations ready; those generals who made use of them did not lose one battle in ten; those who ran counter to them in any particular saw their armies incontinently beaten and put to flight." Ts'ao Kung's notes on Sun Tzŭ, models of austere brevity, are so thoroughly characteristic of the stern commander known to history, that it is hard indeed to conceive of them as the work of a mere *littérateur*. Sometimes, indeed, owing to extreme compression, they are scarcely intelligible and stand no less in need of a commentary than the text itself.

Mêng Shih. The commentary which has come down to us under this name is comparatively meager, and nothing about the author is known. Even his personal name has not been recorded. Chi T'ien-pao's edition places him after Chia Lin, and Ch'ao Kung-wu also assigns him to the T'ang dynasty, but this is a mistake. In Sun Hsing-yen's preface, he appears as Mêng Shih of the Liang dynasty [502–557]. Others would identify him with Mêng K'ang of the 3rd century. He is named in one work as the last of the "Five Commentators," the others being Wei Wu Ti, Tu Mu, Ch'ên Hao and Chia Lin.

Li Ch'üan of the 8th century was a well-known writer on military tactics. [One of his works] has been in constant use down to the present day. [Another work] mentions (lives of famous generals from the Chou to the T'ang dynasty) as written by him. According to Ch'ao Kung-wu and the *T'ien-i-ko* catalogue, he followed a variant of the text of Sun Tzŭ which differs considerably from those now extant. His notes are mostly short and to the point, and he frequently illustrates his remarks by anecdotes from Chinese history.

Tu Yu (died 812) did not publish a separate commentary on Sun Tzŭ, his notes being taken from the *T'ung Tien*, the encyclopedic treatise on the Constitution which was his life-work. They are largely repetitions

兵
法

of Ts'ao Kung and Mêng Shih, besides which it is believed that he drew on the ancient commentaries of Wang Ling and others. Owing to the peculiar arrangement of *T'ung Tien*, he has to explain each passage on its merits, apart from the context, and sometimes his own explanation does not agree with that of Ts'ao Kung, whom he always quotes first. Though not strictly to be reckoned as one of the "Ten Commentators," he was added to their number by Chi T'ien-pao, being wrongly placed after his grandson Tu Mu.

Tu Mu (803–852) is perhaps best known as a poet—a bright star even in the glorious galaxy of the T'ang period. We learn from Ch'ao Kung-wu that although he had no practical experience of war, he was extremely fond of discussing the subject, and was moreover well read in the military history of the *Ch'un Ch'iu* and *Chan Kuo* eras. His notes, therefore, are well worth attention. They are very copious, and replete with historical parallels. The gist of Sun Tzŭ's work is thus summarized by him: "Practice benevolence and justice, but on the other hand make full use of artifice and measures of expediency." He further declared that all the military triumphs and disasters of the thousand years which had elapsed since Sun Tzŭ's death would, upon examination, be found to uphold and corroborate, in every particular, the maxims contained in his book. Tu Mu's somewhat spiteful charge against Ts'ao Kung has already been considered elsewhere.

Ch'ên Hao appears to have been a contemporary of Tu Mu. Ch'ao Kung-wu says that he was impelled to write a new commentary on Sun Tzŭ because Ts'ao Kung's on the one hand was too obscure and subtle, and that of Tu Mu on the other too long-winded and diffuse. Ou-yang Hsiu, writing in the middle of the 11th century, calls Ts'ao Kung, Tu Mu and Ch'ên Hao the three chief commentators on Sun Tzŭ, and

孫
子

observes that Ch'ên Hao is continually attacking Tu Mu's shortcomings. His commentary, though not lacking in merit, must rank below those of his predecessors.

Chia Lin is known to have lived under the T'ang dynasty, for his commentary on Sun Tzŭ is mentioned in the *T'ang Shu* and was afterwards republished by Chi Hsieh of the same dynasty together with those of Mêng Shih and Tu Yu. It is of somewhat scanty texture, and in point of quality, too, perhaps the least valuable of the eleven.

Mei Yao-ch'ên (1002–1060), commonly known by his "style" as Mei Shêng-yü, was, like Tu Mu, a poet of distinction. His commentary was published with a laudatory preface by the great Ou-yang Hsiu, from which we may cull the following:

> Later scholars have misread Sun Tzŭ, distorting his words and trying to make them square with their own one-sided views. Thus, though commentators have not been lacking, only a few have proved equal to the task. My friend Shêng-yü has not fallen into this mistake. In attempting to provide a critical commentary for Sun Tzŭ's work, he does not lose sight of the fact that these sayings were intended for states engaged in internecine warfare; that the author is not concerned with the military conditions prevailing under the sovereigns of the three ancient dynasties, nor with the nine punitive measures prescribed to the Minister of War. Again, Sun Wu loved brevity of diction, but his meaning is always deep. Whether the subject be marching an army, or handling soldiers, or estimating the enemy, or controlling the forces of victory, it is always systematically treated; the sayings are bound together in strict logical sequence, though this has

孫
子

been obscured by commentators who have probably failed to grasp their meaning. In his own commentary, Mei Shêng-yü has brushed aside all the obstinate prejudices of these critics, and has tried to bring out the true meaning of Sun Tzŭ himself. In this way, the clouds of confusion have been dispersed and the sayings made clear. I am convinced that the present work deserves to be handed down side by side with the three great commentaries; and for a great deal that they find in the sayings, coming generations will have constant reason to thank my friend Shêng-yü.

Making some allowance for the exuberance of friendship, I am inclined to endorse this favorable judgment, and would certainly place him above Ch'ên Hao in order of merit.

Wang Hsi, also of the Sung dynasty, is decidedly original in some of his interpretations, but much less judicious than Mei Yao-ch'ên, and on the whole not a very trustworthy guide. He is fond of comparing his own commentary with that of Ts'ao Kung, but the comparison is not often flattering to him. We learn from Ch'ao Kung-wu that Wang Hsi revised the ancient text of Sun Tzŭ, filling up lacunae and correcting mistakes.

Ho Yen-hsi of the Sung dynasty. The personal name of this commentator is given as above by Chêng Ch'iao in the *T'ung Chih*, written about the middle of the twelfth century, but he appears simply as Ho Shih in the *Yu Hai*, and Ma Tuan-lin quotes Ch'ao Kung-wu as saying that his personal name is unknown. There seems to be no reason to doubt Chêng Ch'iao's statement, otherwise I should have been inclined to hazard a guess and identify him with one Ho Ch'ü-fei, the author of a short treatise on war, who lived in the latter part of the 11th century.

Ho Shih's commentary, in the words of the *T'ien-i-ko* catalogue, "contains helpful additions" here and there, but is chiefly remarkable for the copious extracts taken, in adapted form, from the dynastic histories and other sources.

Chang Yü. The list closes with a commentator of no great originality perhaps, but gifted with admirable powers of lucid exposition. His commentator is based on that of Ts'ao Kung, whose terse sentences he contrives to expand and develop in masterly fashion. Without Chang Yu, it is safe to say that much of Ts'ao Kung's commentary would have remained cloaked in its pristine obscurity and therefore valueless. His work is not mentioned in the Sung history, the *T'ung K'ao*, or the *Yü Hai*, but it finds a niche in the *T'ung Chih*, which also names him as the author of the "Lives of Famous Generals."

It is rather remarkable that the last-named four should all have flourished within so short a space of time. Ch'ao Kung-wu accounts for it by saying: "During the early years of the Sung dynasty the Empire enjoyed a long spell of peace, and men ceased to practice the art of war. But when (Chao) Yüan-hao's rebellion came [1038–1042] and the frontier generals were defeated time after time, the Court made strenuous inquiry for men skilled in war, and military topics became the vogue amongst all the high officials. Hence it is that the commentators of Sun Tzŭ in our dynasty belong mainly to that period."

Besides these eleven commentators, there are several others whose work has not come down to us. The *Sui Shu* mentions four, namely Wang Ling (often quoted by Tu Yu as Wang Tzŭ); Chang Tzŭ-shang; Chia Hsü of Wei; and Shên Yu of Wu. The *T'ang Shu* adds Sun Hao, and the *T'ung Chih Hsiao Chi*, while the *T'u Shu* mentions a Ming commentator, Huang Jun-yu. It is possible that some of these may have been merely collectors and editors of other commentaries, like Chi T'ien-pao and Chi Hsieh, mentioned above. [Certainly in the case of the latter, the entry in the *T'ung K'ao*, without the following note, would give one to understand that he had written an independent commentary of his own.]

There are two works, described in the *Ssu K'u Ch'üan Shu* and no doubt extremely rare, which I should much like to have seen. [One appears] in 5 *chüan*. It gives selections from four new commentators, probably of the Ming dynasty, as well as from the eleven known to us. The names of the four are Hsieh Yüan; Chang Ao; Li Ts'ai; and Huang Chih-chêng. The other work [appears] in 4 *chüan*, compiled by Chêng Tuan of the present dynasty. It is a compendium of information on Sun Tzŭ's 13 chapters.

APPRECIATIONS OF SUN TZŬ

Sun Tzŭ has exercised a potent fascination over the minds of some of China's greatest men. Among the famous generals who are known to have studied his pages with enthusiasm may be mentioned Han Hsin (*d.* 196 BC), Fêng I (*d.* AD 34), Lü Mêng (*d.* 219), and Yo Fei (1103–1141). The opinion of Ts'ao Kung, who disputes with Han Hsin the highest place in Chinese military annals, has already been recorded. Still more remarkable, in one way, is the testimony of purely literary men, such as Su Hsün (the father of Su Tung-p'o), who wrote several essays on military topics, all of which owe their chief inspiration to Sun Tzŭ. The following short passage by him is preserved in the *Yü Hai*:

> Sun Wu's saying, that in war one cannot make
> certain of conquering, is very different indeed from
> what other books tell us. Wu Ch'i was a man of the
> same stamp as Sun Wu: they both wrote books on

war, and they are linked together in popular speech as "Sun and Wu." But Wu Ch'i's remarks on war are less weighty, his rules are rougher and more crudely stated, and there is not the same unity of plan as in Sun Tzŭ's work, where the style is terse, but the meaning fully brought out.

[The following is an extract from the "Impartial Judgments in the Garden of Literature" by Chêng Hou]:

> Sun Tzŭ's 13 chapters are not only the staple and base of all military men's training, but also compel the most careful attention of scholars and men of letters. His sayings are terse yet elegant, simple yet profound, perspicuous and eminently practical. Such works as the *Lun Yü*, the *I Ching* and the great Commentary, as well as the writings of Mencius, Hsün K'uang and Yang Chu, all fall below the level of Sun Tzŭ.

孫
子

Chu Hsi, commenting on this, fully admits the first part of the criticism, although he dislikes the audacious comparison with the venerated classical works. Language of this sort, he says, "encourages a ruler's bent towards unrelenting warfare and reckless militarism."

APOLOGIES FOR WAR

Accustomed as we are to think of China as the greatest peace-loving nation on earth, we are in some danger of forgetting that her experience of war in all its phases has also been such as no modern State can parallel. Her long military annals stretch back to a point at which they are lost in the mists of time. She had built the Great Wall and was maintaining a huge standing army along her frontier centuries before the first Roman legionary was seen on the Danube. What with the perpetual collisions of the ancient feudal States, the grim conflicts with Huns, Turks and other invaders after the centralization of government, the terrific upheavals which accompanied the overthrow of so many dynasties, besides the countless rebellions and minor disturbances that have flamed up and flickered out again one by one, it is hardly too much to say that the clash of arms has never ceased to resound in one portion or another of the Empire.

No less remarkable is the succession of illustrious captains to whom China can point with pride. As in all countries, the greatest are fond of emerging at the most fateful crises of her history. Thus, Po Ch'i stands out conspicuous in the period when Ch'in was entering upon her final struggle with the remaining independent states. The stormy years which followed the break-up of the Ch'in dynasty are illuminated by the transcendent genius of Han Hsin. When the House of Han in turn is tottering to its fall, the great and baleful figure of Ts'ao Ts'ao dominates the scene. And in the establishment of the T'ang dynasty, one of the mightiest tasks achieved by man, the superhuman energy of Li Shihmin (afterwards the Emperor T'ai Tsung) was seconded by the brilliant strategy of Li Ching. None of these generals need fear comparison with the greatest names in the military history of Europe.

In spite of all this, the great body of Chinese sentiment, from Lao Tzŭ downwards, and especially as reflected in the standard literature of Confucianism, has been consistently pacific and intensely opposed to militarism in any form. It is such an uncommon thing to find any of the literati defending warfare on principle, that I have thought it worth while to collect and translate a few passages in which the unorthodox view is upheld. The following, by Ssŭ-ma Ch'ien, shows that for all his ardent admiration of Confucius, he was yet no advocate of peace at any price:

> Military weapons are the means used by the Sage
> to punish violence and cruelty, to give peace to trou-

(continued on page 62)

blous times, to remove difficulties and dangers, and to succor those who are in peril. Every animal with blood in its veins and horns on its head will fight when it is attacked. How much more so will man, who carries in his breast the faculties of love and hatred, joy and anger! When he is pleased, a feeling of affection springs up within him; when angry, his poisoned sting is brought into play. That is the natural law which governs his being. . . . What then shall be said of those scholars of our time, blind to all great issues, and without any appreciation of relative values, who can only bark out their stale formulas about "virtue" and "civilization," condemning the use of military weapons? They will surely bring our country to impotence and dishonor and the loss of her rightful heritage; or, at the very least, they will bring about invasion and rebellion, sacrifice of territory and general enfeeblement. Yet they obstinately refuse to modify the position they have taken up. The truth is that, just as in the family the teacher must not spare the rod, and punishments cannot be dispensed with in the State, so military chastisement can never be allowed to fall into abeyance in the Empire. All one can say is that this power will be exercised wisely by some, foolishly by others, and that among those who bear arms some will be loyal and others rebellious.

The next piece is taken from Tu Mu's preface to his commentary on Sun Tzŭ:

War may be defined as punishment, which is one of the functions of government. It was the profession of Chung Yu and Jan Ch'iu, both disciples of Confucius. Nowadays, the holding of trials and hearing of litigation, the imprisonment of offenders and their execution by flogging in the market-place, are all done by officials. But the wielding of huge armies, the throwing down of fortified cities, the hauling of women and children into captivity, and the beheading of traitors—this is also work which is done by officials. The objects of the rack and of military weapons are essentially the same. There is no intrinsic difference between the punishment of flogging and cutting off heads in war. For the lesser infractions of law, which are easily dealt with, only a small amount of force need be employed: hence the use of military weapons and wholesale decapitation. In both cases, however, the end in view is to get rid of wicked people, and to give comfort and relief to the good. . . .

Chi-sun asked Jan Yu, saying: "Have you, Sir, acquired your military aptitude by study, or is it innate?" Jan Yu replied: "It has been acquired by study." "How can that be so," said Chi-sun, "seeing that you are a disciple of Confucius?" "It is a fact," replied Jan Yu; "I was taught by Confucius. It is fitting that the great Sage should exercise both civil and military functions, though to be sure my instruction in the art of fighting has not yet gone very far."

兵
法

Now, who the author was of this rigid distinction between the "civil" and the "military," and the limitation of each to a separate sphere of action, or in what year of which dynasty it was first introduced, is more than I can say. But, at any rate, it has come about that the members of the governing class are quite afraid of enlarging on military topics, or do so only in a shamefaced manner. If any are bold enough to discuss the subject, they are at once set down as eccentric individuals of coarse and brutal propensities.

This is an extraordinary instance in which, through sheer lack of reasoning, men unhappily lose sight of fundamental principles.

When the Duke of Chou was minister under Ch'êng Wang, he regulated ceremonies and made music, and venerated the arts of scholarship and learning; yet when the barbarians of the River Huai revolted, he sallied forth and chastised them. When Confucius held office under the Duke of Lu, and a meeting was convened at Chia-ku, he said: "If pacific negotiations are in progress, warlike preparations should have been made beforehand." He rebuked and shamed the Marquis of Ch'i, who cowered under him and dared not proceed to violence. How can it be said that these two great Sages had no knowledge of military matters?

孫
子

We have seen that the great Chu Hsi held Sun Tzǔ in high esteem. He also appeals to the authority of the Classics:

Our Master Confucius, answering Duke Ling of Wei, said: "I have never studied matters connected with armies and battalions." Replying to K'ung Wên-tzuǔ, he said: "I have not been instructed about buff-coats and weapons." But if we turn to the meeting at Chia-ku, we find that he used armed force against the men of Lai, so that the marquis of Ch'i was over-awed. Again, when the inhabitants of Pi revolted, the ordered his officers to attack them, whereupon they were defeated and fled in confusion. He once uttered the words: "If I fight, I conquer." And Jan Yu also said: "The Sage exercises both civil and military functions." Can it be a fact that Confucius never studied or received instruction in the art of war? We can only say that he did not specially choose matters connected with armies and fighting to be the subject of his teaching.

Sun Hsing-yen, the editor of Sun Tzŭ, writes in similar strain:

> Confucius said: "I am unversed in military matters." He also said: "If I fight, I conquer." Confucius ordered ceremonies and regulated music. Now war constitutes one of the five classes of State ceremonial, and must not be treated as an independent branch of study. Hence, the words "I am unversed in" must be taken to mean that there are things which even an inspired Teacher does not know. Those who have to lead an army and devise stratagems, must learn the art of war. But if one can command the services of a good general like Sun Tzŭ, who was employed by Wu Tzŭ-hsü, there is no need to learn it oneself. Hence the remark added by Confucius: "If I fight, I conquer."
>
> The men of the present day, however, willfully interpret these words of Confucius in their narrowest sense, as though he meant that books on the art of war were not worth reading. With blind persistency, they adduce the example of Chao Kua, who pored over his father's books to no purpose, as a proof that all military theory is useless. Again, seeing that books on war have to do with such things as opportunism in designing plans, and the conversion of spies, they hold that the art is immoral and unworthy of a sage. These people ignore the fact that the studies of our scholars and the civil administration of our officials also require steady application and practice before efficiency is reached. The ancients were particularly chary of allowing mere novices to botch their work. Weapons are baneful and fighting perilous; and use-

兵
法

less unless a general is in constant practice, he ought not to hazard other men's lives in battle. Hence it is essential that Sun Tzŭ's 13 chapters should be studied. Hsiang Liang used to instruct his nephew Chi in the art of war. Chi got a rough idea of the art in its general bearings, but would not pursue his studies to their proper outcome, the consequence being that he was finally defeated and overthrown. He did not realize that the tricks and artifices of war are beyond verbal computation. Duke Hsiang of Sung and King Yen of Hsü were brought to destruction by their misplaced humanity. The treacherous and underhand nature of war necessitates the use of guile and stratagem suited to the occasion. There is a case on record of Confucius himself having violated an extorted oath, and also of his having left the Sung State in disguise. Can we then recklessly arraign Sun Tzŭ for disregarding truth and honesty?

孫
子

BIBLIOGRAPHY

The following are the oldest Chinese treatises on war, after Sun Tzŭ. The notes on each have been drawn principally from the *Ssu k'u ch'üan shu chien ming mu lu*, ch 9, fol. 22 s99.

Wu Tzŭ, in 1 *chüan* or 6 chapters. By Wu Ch'i (*d.* 381 BC). A genuine work. See *Shih Chi.*

Ssŭ-ma Fa, in 1 *chüan* or 5 chapters. Wrongly attributed to Ssŭ-ma Jang-chü of the 6th century BC. Its date, however, must be early, as the customs of the three ancient dynasties are constantly to be met within its pages. See *Shih Chi*, ch. 64.

The *Ssŭ K'u Ch'üan Shu* remarks that the oldest three treatises on war—Sun Tzŭ, *Wu Tzŭ* and *Ssŭ-ma Fa*, are, generally speaking, only concerned with things strictly military—the art of producing, collecting, training and drilling troops, and the correct theory with regard to measures of expediency, laying plans, transport of goods and the handling of soldiers—in strong contrast to later works, in which the science of war is usually blended with metaphysics, divination and magical arts in general.

Liu T'ao, in 6 *chüan*, or 60 chapters. Attributed to Lü Wang (or Lü Shang, also known as T'ai Kung) of the 12th century BC. But its style does not belong to the era of the Three Dynasties. Lu Tê-ming (AD 550–625) mentions the work, and enumerates the headings of the six sections so that the forgery cannot have been later than Sui dynasty.

Wei Liao Tzŭ, in 5 *chüan*. Attributed to Wei Liao (4th cent. BC), who studied under the famous Kuei-ku Tzŭ. The work appears to have been originally in 31 chapters, whereas the text we possess contains only 24. Its matter is sound enough in the main, though the strategical devices differ considerably from those of the Warring States period. It has been furnished with a commentary by the well-known Sung philosopher Chang Tsai.

兵
法

San Lüeh, in 3 *chüan*. Attributed to Huang-shih Kung, a legendary personage who is said to have bestowed it on Chang Liang (*d*. 187 BC) in an interview on a bridge. But here again, the style is not that of works dating from the Ch'in or Han period. The Han Emperor Kuang Wu (AD 25–57) apparently quotes from it in one of his proclamations; but the passage in question may have been inserted later on, in order to prove the genuineness of the work. We shall not be far out if we refer it to the Northern Sung period (AD 420–478), or somewhat earlier.

Li Wei Kung Wên Tui, in 3 sections. Written in the form of a dialogue between T'ai Tsung and his great general Li Ching, it is usually ascribed to the latter. Competent authorities consider it a forgery, though the author was evidently well versed in the art of war.

Li Ching Ping Fa (not to be confounded with the foregoing) is a short treatise in 8 chapters, preserved in the T'ung Tien, but not published separately. This fact explains its omission from the *Ssŭ K'u Ch'üan Shu*.

Wu Ch'i Ching, in 1 *chüan*. Attributed to the legendary minister Feng Hou, with exegetical notes by Kung-sun Hung of the Han dynasty (*d.* 121 BC), and said to have been eulogized by the celebrated general Ma Lung (*d.* AD 300). Yet the earliest mention of it is in the *Sung Chih*. Although a forgery, the work is well put together.

Considering the high popular estimation in which Chu-ko Liang has always been held, it is not surprising to find more than one work on war ascribed to his pen. Such are (1) the *Shih Liu Ts'ê* (1 chüan), preserved in the *Yung Lo Ta Tien;* (2) *Chiang Yüan* (1 chüan); and (3) *Hsin Shu* (1 chüan), which steals wholesale from Sun Tzŭ. None of these has the slightest claim to be considered genuine. Most of the large Chinese encyclopedias contain extensive sections devoted to the literature of war. The following references may be found useful:

T'ung Tien (circa AD 800), ch. 148–162.

T'ai P'ing Yü Lan (983), ch. 270–359.

Wên Hsien T'ung K'ao (13th cent.), ch. 221.

Yü Hai (13th cent.), ch. 140, 141.

San Ts'ai T'u Hui (16th cent).

Kuang Po Wu Chih (1607), ch. 31, 32.

Ch'ien Ch'io Lei Shu (1632), ch. 75.

Yüan Chien Lei Han (1710), ch. 206–229.

Ku Chin T'u Shu Chi Ch'êng (1726), section XXX, esp. chs. 81–90.

Hsü Wên Hsien T'ung K'ao (1784), ch. 121–134.

Huang Ch'ao Ching Shih Wên Pien (1826), ch. 76, 77.

The bibliographical sections of certain historical works also deserve mention:

Ch'ien Han Shu, ch. 30.

Sui Shu, ch. 32–35.

Chiu T'ang Shu, ch. 46, 47.

Hsin T'ang Shu, ch. 57, 60.

Sung Shih, ch. 202–209.

T'ung Chih (circa 1150), ch. 68.

兵
法

To these of course must be added the great Catalogue of the Imperial Library:

Ssu K'uŭ Ch'üan Shu Tsung Mu T'i Yao (1790), ch. 99, 100.

LAYING PLANS

[Earlier translators have incorrectly titled this chapter "First Principles."] Ts'ao Kung says it refers to the deliberations in the temple selected by the general for his temporary use, or as we should say, in his tent. *See sect. 26.*

孫
子

1. Sun Tzŭ said: The art of war is of vital importance to the State.

2. It is a matter of life and death, a road either to safety or to ruin. Hence it is a subject of inquiry which can on no account be neglected.

3. The art of war, then, is governed by five constant factors, to be taken into account in one's deliberations, when seeking to determine the conditions obtaining in the field.

4. These are: (1) The Moral Law; (2) Heaven; (3) Earth; (4) The Commander; (5) Method and discipline.

   It appears from what follows that Sun Tzŭ means by ["Moral Law"] a principle of harmony, not unlike the Tao of Lao Tzŭ in its moral aspect. One might be tempted to render it by "morale," were it not considered as an attribute of the ruler in sect. 13.

5, 6. *The Moral Law* causes the people to be in complete accord with their ruler, so that they will follow him regardless of their lives, undismayed by any danger.

7. *Heaven* signifies night and day, cold and heat, times and seasons.

   The commentators, I think, make an unnecessary mystery of [the wording here]. Thus Mêng Shih defines the words as "the hard and the soft, waxing and waning" of Heaven. Wang Hsi, however, may be right in saying that what is meant is "the general economy of Heaven," including the five elements, the four seasons, wind and clouds, and other phenomena.

8. *Earth* comprises distances, great and small; danger and security; open ground and narrow passes; the chances of life and death.

9. *The Commander* stands for the virtues of wisdom, sincerity, benevolence, courage and strictness.

The five cardinal virtues of the Chinese are (1) humanity or benevolence; (2) uprightness of mind; (3) self-respect, self-control, or "proper feeling"; (4) wisdom; (5) sincerity or good faith. Here [wisdom] and [sincerity or good faith] are put before [humanity or benevolence], and the two military virtues of "courage" and "strictness" substituted for [uprightness of mind] and [self-respect, self-control, or "proper feeling"].

兵
法

10. By *Method and discipline* are to be understood the marshaling of the army in its proper subdivisions, the graduations of rank among the officers, the maintenance of roads by which supplies may reach the army, and the control of military expenditure.

11. These five heads should be familiar to every general: he who knows them will be victorious; he who knows them not will fail.

12. Therefore, in your deliberations, when seeking to determine the military conditions, let them be made the basis of a comparison, in this wise:

13. (1) Which of the two sovereigns is imbued with the Moral law?

I.e., "is in harmony with his subjects." See sect. 5.

(2) Which of the two generals has most ability?

(3) With whom lie the advantages derived from Heaven and Earth?

See sects. 7, 8.

(4) On which side is discipline most rigorously enforced?

Tu Mu alludes to the remarkable story of Ts'ao Ts'ao (AD 155–220), who was such a strict disciplinarian that once, in accordance with his own severe regulations against injury to standing crops, he condemned himself to death for having allowed his horse to shy into a field of corn! However, in lieu of losing his head, he was persuaded to satisfy his sense of justice by cutting off his hair. Ts'ao Ts'ao's own comment on the present passage is characteristically curt: "when you lay down a law, see that it is not disobeyed; if it is disobeyed the offender must be put to death."

(5) Which army is stronger?

Morally as well as physically. As Mei Yao-ch'ên puts it, [freely rendered], *"esprit de corps* and 'big battalions.'"

(6) On which side are officers and men more highly trained?

Tu Yu quotes [Wang Tzǔ] as saying: "Without constant practice, the officers will be nervous and undecided when mustering for battle; without constant practice, the general will be wavering and irresolute when the crisis is at hand."

(7) In which army is there the greater constancy both in reward and punishment?

[Literally, "clear"]; that is, on which side is there the most absolute certainty that merit will be properly rewarded and misdeeds summarily punished?

14. **By means of these seven considerations I can forecast victory or defeat.**

15. **The general that hearkens to my counsel and acts upon it, will conquer:—let such a one be retained in command! The general that hearkens not to my counsel nor acts upon it, will suffer defeat:—let such a one be dismissed!**

The form of this paragraph reminds us that Sun Tzŭ's treatise was composed expressly for the benefit of his patron Ho Lü, king of the Wu State.

16. **While heeding the profit of my counsel, avail yourself also of any helpful circumstances over and beyond the ordinary rules.**

兵
法

17. **According as circumstances are favorable, one should modify one's plans.**

Sun Tzŭ, as a practical soldier, will have none of the "bookish theorie." He cautions us here not to pin our faith to abstract principles; "for," as Chang Yü puts it, "while the main laws of strategy can be stated clearly enough for the benefit of all and sundry, you must be guided by the actions of the enemy in attempting to secure a favorable position in actual warfare." On the eve of the battle of Waterloo, Lord Uxbridge, commanding the cavalry, went to the Duke of Wellington in order to learn what his plans and calculations were for the morrow, because, as he explained, he might suddenly find himself Commander-in-chief and would be unable to frame new plans in a critical moment. The Duke listened quietly and then said: "Who will attack the first tomorrow—I or Bonaparte?" "Bonaparte," replied Lord Uxbridge. "Well," continued the Duke, "Bonaparte has not given me any idea of his projects; and as my plans will depend upon his, how can you expect me to tell you what mine are?" (*Words on Wellington*, Sir. W. Fraser.)

18. **All warfare is based on deception.**

    The truth of this pithy and profound saying will be admitted by every soldier. Col. Henderson tells us that Wellington, great in so many military qualities, was especially distinguished by "the extraordinary skill with which he concealed his movements and deceived both friend and foe."

19. **Hence, when able to attack, we must seem unable; when using our forces, we must seem inactive; when we are near, we must make the enemy believe we are far away; when far away, we must make him believe we are near.**

20. **Hold out baits to entice the enemy. Feign disorder, and crush him.**

    [All commentators, except Chang Yü say, of the enemy]: "when he is in disorder, crush him." It is more natural to suppose that Sun Tzŭ is still illustrating the uses of deception in war.

21. **If he is secure at all points, be prepared for him. If he is in superior strength, evade him.**

22. **If your opponent is of choleric temper, seek to irritate him. Pretend to be weak, that he may grow arrogant.**

    Wang Tzŭ, quoted by Tu Yu, says that the good tactician plays with his adversary as a cat plays with a mouse, first feigning weakness and immobility, and then suddenly pouncing upon him.

23. **If he is taking his ease, give him no rest.**

This is probably the meaning, though Mei Yao-ch'ên has the note: "while we are taking our ease, wait for the enemy to tire himself out." The *Yu Lan* has "Lure him on and tire him out."

If his forces are united, separate them.

Less plausible is the interpretation favored by most of the commentators: "If sovereign and subject are in accord, put division between them."

24. **Attack him where he is unprepared, appear where you are not expected.**

25. **These military devices, leading to victory, must not be divulged beforehand.**

    This seems to be the way in which Ts'ao Kung understood the passage, and is perhaps the best sense to be gotten out of the text as it stands. Most of the commentators give the following explanation: "It is impossible to lay down rules for warfare before you come into touch with the enemy.". . . As an alternative, however, I would venture to suggest . . . "These maxims for succeeding in war are the first that ought to be imparted."

26. **Now the general who wins a battle makes many calculations in his temple ere the battle is fought.**

    Chang Yü tells us that in ancient times it was customary for a temple to be set apart for the use of a general who was about to take the field, in order that he might there elaborate his plan of campaign.

    The general who loses a battle makes but few calculations beforehand. Thus do many calculations lead to victory, and few calculations to defeat: how much more no calculation at all! It is by attention to this point that I can foresee who is likely to win or lose.

兵
法

WAGING WAR

Ts'ao Kung has the note: "He who wishes to fight must first count the cost," which prepares us for the discovery that the subject of the chapter is not what we might expect from the title, but is primarily a consideration of ways and means.

1. **Sun Tzŭ said: In the operations of war, where there are in the field a thousand swift chariots, as many heavy chariots, and a hundred thousand mail-clad soldiers,**

The [swift chariots] were lightly built and, according to Chang Yü, used for the attack; the "heavy chariots" were heavier, and designed for purposes of defense. Li Ch'üan, it is true, says that the latter were light, but this seems hardly probable. It is interesting to note the analogies between early Chinese warfare and that of the Homeric Greeks. In each case, the war-chariot was the important factor, forming as it did the nucleus round which was grouped a certain number of foot-soldiers. With regard to the numbers given here, we are informed that each swift chariot was accompanied by 75 footmen, and each heavy chariot by 25 footmen, so that the whole army would be divided up into a thousand battalions, each consisting of two chariots and a hundred men.

with provisions enough to carry them a thousand *li*,

2.78 modern *li* go to a mile. The length may have varied slightly since Sun Tzŭ's time.

the expenditure at home and at the front, including entertainment of guests, small items such as glue and paint, and sums spent on chariots and armor, will reach the total of a thousand

ounces of silver per day. Such is the cost of raising an army of 100,000 men.

2. When you engage in actual fighting, if victory is long in coming, then men's weapons will grow dull and their ardor will be damped. If you lay siege to a town, you will exhaust your strength.

3. Again, if the campaign is protracted, the resources of the State will not be equal to the strain.

4. Now, when your weapons are dulled, your ardor damped, your strength exhausted and your treasure spent, other chieftains will spring up to take advantage of your extremity. Then no man, however wise, will be able to avert the consequences that must ensue.

兵
法

5. Thus, though we have heard of stupid haste in war, cleverness has never been seen associated with long delays.

This concise and difficult sentence is not well explained by any of the commentators. Ts'ao Kung, Li Ch'üan, Mêng Shih, Tu Yu, Tu Mu and Mei Yao-ch'ên have notes to the effect that a general, though naturally stupid, may nevertheless conquer through sheer force of rapidity. Ho Shih says: "Haste may be stupid, but at any rate it saves expenditure of energy and treasure; protracted operations may be very clever, but they bring calamity in their train." Wang Hsi evades the difficulty by remarking: "Lengthy operations mean an army growing old, wealth being expended, an empty exchequer and distress among the people; true cleverness insures against the occurrence of such calamities." Chang Yü says: "So long as victory can be attained, stupid haste is preferable to clever dilatoriness." Now Sun

(continued on page 86)

Tzŭ says nothing whatever, except possibly by implication, about ill-considered haste being better than ingenious but lengthy operations. What he does say is something much more guarded, namely that, while speed may sometimes be injudicious, tardiness can never be anything but foolish—if only because it means impoverishment to the nation. . . . In considering the point raised here by Sun Tzŭ, the classic example of Fabius Cunctator will inevitably occur to the mind. That general deliberately measured the endurance of Rome against that of Hannibal's isolated army, because it seemed to him that the latter was more likely to suffer from a long campaign in a strange country. But it is quite a moot question whether his tactics would have proved successful in the long run. Their reversal it is true, led to Cannae; but this only establishes a negative presumption in their favor.

6. **There is no instance of a country having benefited from prolonged warfare.**

7. **It is only one who is thoroughly acquainted with the evils of war that can thoroughly understand the profitable way of carrying it on.**

   That is, with rapidity. Only one who knows the disastrous effects of a long war can realize the supreme importance of rapidity in bringing it to a close. Only two commentators seem to favor this interpretation, but it fits well into the logic of the context, whereas the rendering, "He who does not know the evils of war cannot appreciate its benefits," is distinctly pointless.

8. **The skillful soldier does not raise a second levy, neither are his supply-wagons loaded more than twice.**

   Once war is declared, he will not waste precious time in waiting for reinforcements, nor will he return his army back for fresh supplies, but crosses the enemy's frontier without delay. This may seem an audacious policy to recommend, but with all great strategists, from Julius Caesar to Napoleon Bonaparte, the value of time—that is, being a little ahead of your opponent—has counted for more than either numerical superiority or the nicest calculations with regard to commissariat.

9. **Bring war material with you from home, but forage on the enemy. Thus the army will have food enough for its needs.**

[The Chinese word interpreted as "war material" translates as] "things to be used," in the widest sense. It includes all the impedimenta of an army, apart from provisions.

10. **Poverty of the State exchequer causes an army to be maintained by contributions from a distance. Contributing to maintain an army at a distance causes the people to be impoverished.**

The beginning of this sentence does not balance properly with the next, though obviously intended to do so. The arrangement, moreover, is so awkward that I cannot help suspecting some corruption in the text. It never seems to occur to Chinese commentators that an emendation may be necessary for the sense, and we get no help from them there. [The words Sun Tzŭ used to indicate the cause of the people's impoverishment] have reference to some system by which the husbandmen sent their contributions of corn to the army direct.

11. **On the other hand, the proximity of an army causes prices to go up; and high prices cause the people's substance to be drained away.**

[Wang Hsi interprets the first part of this section to mean that high prices occur] before the army has left its own territory. Ts'ao Kung understands it of an army that has already crossed the frontier.

兵
法

12. When their substance is drained away, the peasantry will be afflicted by heavy exactions.

13, 14. With this loss of substance and exhaustion of strength, the homes of the people will be stripped bare, and three-tenths of their income will be dissipated;

[Tu Mu and Wang Hsi agree that part of this section means] the people are mulcted not of 3/10, but of 7/10, of their income. . . . Ho Shih has a characteristic tag: "The *people* being regarded as the essential part of the State, and *food* as the people's heaven, is it not right that those in authority should value and be careful of both?"

while Government expenses for broken chariots, worn-out horses, breast-plates and helmets, bows and arrows, spears and shields, protective mantlets, draft-oxen and heavy wagons, will amount to four-tenths of its total revenue.

15. Hence a wise general makes a point of foraging on the enemy. One cartload of the enemy's provisions is equivalent to twenty of one's own, and likewise a single picul of his provender is equivalent to twenty from one's own store.

Because twenty cartloads will be consumed in the process of transporting one cartload to the front.

16. Now in order to kill the enemy, our men must be roused to anger; that there may be advantage from defeating the enemy, they must have their rewards.

Tu Mu says: "Rewards are necessary in order to make the soldiers see the advantage of beating the enemy; thus, when you capture spoils from the enemy, they must be used as rewards, so that all your men may have a keen desire to fight, each on his own account."

17. Therefore in chariot fighting, when ten or more chariots have been taken, those should be rewarded who took the first. Our own flags should be substituted for those of the enemy, and the chariots mingled and used in conjunction with ours. The captured soldiers should be kindly treated and kept.

18. This is called using the conquered foe to augment one's own strength.

19. In war, then, let your great object be victory, not lengthy campaigns.

    As Ho Shih remarks: "War is not a thing to be trifled with." Sun Tzŭ here reiterates the main lesson which this chapter is intended to enforce.

20. Thus it may be known that the leader of armies is the arbiter of the people's fate, the man on whom it depends whether the nation shall be in peace or in peril.

ATTACK BY STRATAGEM

1. Sun Tzŭ said: In the practical art of war, the best thing of all is to take the enemy's country whole and intact; to shatter and destroy it is not so good. So, too, it is better to recapture an army entire than to destroy it, to capture a regiment, a detachment or a company entire than to destroy them.

[An] army corps, according to the Ssŭ-ma Fa, consisted nominally of 12,500 men; according to Ts'ao Kung, [the equivalent of a regiment] contained 500 men, [the equivalent of a detachment] consists from any number between 100 and 500, and [the equivalent of a company] contains from 5 to 100 men. For the last two, however, Chang Yü gives the exact figures of 100 and 5 respectively.

2. Hence to fight and conquer in all your battles is not supreme excellence; supreme excellence consists in breaking the enemy's resistance without fighting.

Here again, no modern strategist but will approve the words of the old Chinese general. Moltke's greatest triumph, the capitulation of the huge French army at Sedan, was won practically without bloodshed.

3. Thus the highest form of generalship is to balk the enemy's plans;

Perhaps the word "balk" falls short of expressing the full force of [the wording], which implies not an attitude of defense, whereby one might be content to foil the enemy's stratagems one after another, but an active policy of counter-attack. Ho Shih puts this very clearly in his note: "When the enemy has made a plan of attack against us, we must anticipate him by delivering our own attack first."

the next best is to prevent the junction of the enemy's forces;

Isolating him from his allies. We must not forget that Sun Tzŭ, in speaking of hostilities, always has in mind the numerous states or principalities into which the China of his day was split up.

the next in order is to attack the enemy's army in the field;

When he is already at full strength.

and the worst policy of all is to besiege walled cities.

4. **The rule is, not to besiege walled cities if it can possibly be avoided.**

Another sound piece of military theory. Had the Boers acted upon it in 1899, and refrained from dissipating their strength before Kimberley, Mafeking, or even Ladysmith, it is more than probable that they would have been masters of the situation before the British were ready seriously to oppose them.

兵
法

The preparation of mantlets, movable shelters, and various implements of war, will take up three whole months;

It is not quite clear what [the word, here translated as "mantlets"] were. Ts'ao Kung simply defines them as "large shields," but we get a better idea of them from Li Ch'üan, who says they were to protect the heads of those who were assaulting the city walls at close quarters. This seems to suggest a sort of Roman *testudo*, ready made. Tu Mu says they were . . . (wheeled vehicles used in repelling attack, according to K'ang His), but this is denied by Ch'ên Hao. See *ch.* II, *sect.* 14. The name is also applied to turrets on city walls. Of the [movable shelters] we get a fairly clear description from several commentators. They were wooden missile-proof structures on four wheels, propelled from within, covered over with raw hides, and used in sieges to convey parties of men to and from the walls, for the purpose of filling up the encircling moat with earth. Tu Mu adds that they are now called "wooden donkeys."

and the piling up of mounds over against the walls will take three months more.

[These were] great mounds or ramparts of earth heaped up to the level of the enemy's walls in order to discover the weak points in the defense, and also to destroy the fortified turrets mentioned in the preceding note.

5. The general, unable to control his irritation, will launch his men to the assault like swarming ants,

This vivid simile, . . . as Ts'ao Kung says, is taken from the spectacle of an army of ants climbing a wall. The meaning is that the general, losing patience at the long delay, may make a premature attempt to storm the place before his engines of war are ready.

with the result that one-third of his men are slain, while the town still remains untaken. Such are the disastrous effects of a siege.

6. Therefore the skillful leader subdues the enemy's troops without any fighting; he captures their cities without laying siege to them; he overthrows their kingdom without lengthy operations in the field.

Chia Lin notes that he only overthrows . . . the Government, but does no harm to individuals. The classical instance is Wu Wang, who after having put an end to the Yin dynasty was acclaimed "Father and mother of the people."

7. With his forces intact he will dispute the mastery of the Empire, and thus, without losing a man, his triumph will be complete.

Owing to the double meanings of [the wording], the latter part of the sentence is susceptible of quite a different meaning: "And thus, the weapon not being blunted by use, its keenness remains perfect."

This is the method of attacking by stratagem.

8. **It is the rule in war, if our forces are ten to the enemy's one, to surround him; if five to one, to attack him;**

Straightway, without waiting for any further advantage.

if twice as numerous, to divide our army into two.

Tu Mu takes exception to [this] saying; and at first sight, indeed, it appears to violate a fundamental principle of war. Ts'ao Kung, however, gives a clue to Sun Tzŭ's meaning: "Being two to the enemy's one, we may use one part of our army in the regular way, and the other for some special diversion." . . . Chang Yü thus further elucidates the point: "If our force is twice as numerous as that of the enemy, it should be split up into two divisions, one to meet the enemy in front, and one to fall upon his rear; if he replies to the frontal attack, he may be crushed from behind; if to the rearward attack, he may be crushed in front. This is what is meant by saying that 'one part may be used in the regular way, and the other for some special diversion.' Tu Mu does not understand that dividing one's army is simply an irregular, just as concentrating it is the regular, strategical method, and he is too hasty in calling this a mistake."

9. **If equally matched, we can offer battle;**

Li Ch'üan, followed by Ho Shih, gives the following paraphrase: "If attackers and attacked are equally matched in strength, only the able general will fight."

if slightly inferior in numbers, we can avoid the enemy;

The meaning, "we can *watch* the enemy," is certainly a great improvement on the above; but unfortunately there appears to be no very good authority for the variant. Chang Yü reminds us that the saying only applies if the other factors are equal; a small difference in numbers is often more than counterbalanced by superior energy and discipline.

if quite unequal in every way, we can flee from him.

10. **Hence, though an obstinate fight may be made by a small force, in the end it must be captured by the larger force.**

11. **Now the general is the bulwark of the State; if the bulwark is complete at all points; the State will be strong; if the bulwark is defective, the State will be weak.**

    As Li Ch'üan tersely puts it: "*Ch'i* gap, indicates deficiency; if the general's ability is not perfect (*i.e.*, if he is not thoroughly versed in his profession), his army will lack strength."

12. **There are three ways in which a ruler can bring misfortune upon his army:**

13. **(1) By commanding the army to advance or to retreat, being ignorant of the fact that it cannot obey. This is called hobbling the army.**

    Li Ch'üan . . . adds the comment: "It is like tying together the legs of a thoroughbred, so that it is unable to gallop." One would naturally think of "the ruler" in this passage as being at home, and trying to direct the movements of his army from a distance. But the commentators understand just the reverse, and quote the saying of

T'ai Kung: "A kingdom should not be governed from without, and army should not be directed from within." Of course it is true that, during an engagement, or when in close touch with the enemy, the general should not be in the thick of his own troops, but a little distance apart. Otherwise, he will be liable to misjudge the position as a whole, and give wrong orders.

14. **(2) By attempting to govern an army in the same way as he administers a kingdom, being ignorant of the conditions which obtain in an army. This causes restlessness in the soldier's minds.**

Ts'ao Kung's note . . . may be freely translated: "The military sphere and the civil sphere are wholly distinct; you can't handle an army in kid gloves." And Chang Yü says: "Humanity and justice are the principles on which to govern a state, but not an army; opportunism and flexibility, on the other hand, are military rather than civic virtues."

兵
法

15. **(3) By employing the officers of his army without discrimination,**

That is, he is not careful to use the right man in the right place.

through ignorance of the military principle of adaptation to circumstances. This shakes the confidence of the soldiers.

I follow Mei Yao-ch'ên here. The other commentators [interpret the wording] to refer not to the ruler, as in sects. 13, 14, but to the officers he employs. Thus Tu Yu says: "If a general is ignorant of the principle of adaptability, he must not be entrusted with a position of authority." Tu Mu quotes: "The skillful employer of men will employ the wise man, the brave man, the covetous man, and the stupid man. For the wise man delights in establishing his merit, the brave man likes to show his courage in action, the covetous man is quick at seizing advantages, and the stupid man has no fear of death."

16. But when the army is restless and distrustful, trouble is sure to come from the other feudal princes. This is simply bringing anarchy into the army, and flinging victory away.

17. Thus we may know that there are five essentials for victory:

(1) He will win who knows when to fight and when not to fight.

> Chang Yü says: "If he can fight, he advances and takes the offensive; if he cannot fight, he retreats and remains on the defensive. He will invariably conquer who knows whether it is right to take the offensive or the defensive."

(2) He will win who knows how to handle both superior and inferior forces.

> This is not merely the general's ability to estimate numbers correctly, as Li Ch'üan and others make out. Chang Yü expounds the saying more satisfactorily: "By applying the art of war, it is possible with a lesser force to defeat a greater, and *vice versá*. The secret lies in an eye for locality, and in not letting the right moment slip. Thus Wu Tzŭ says: 'With a superior force, make for easy ground; with an inferior one, make for difficult ground.'"

(3) He will win whose army is animated by the same spirit throughout all its ranks.

(4) He will win who, prepared himself, waits to take the enemy unprepared.

(5) He will win who has military capacity and is not interfered with by the sovereign.

Tu Yu quotes [Wang Tzǔ] as saying: "It is the sovereign's function to give broad instructions, but to decide on battle it is the function of the general." It is needless to dilate on the military disasters which have been caused by undue interference with operations in the field on the part of the home government. Napoleon undoubtedly owed much of his extraordinary success to the fact that he was not hampered by central authority. Victory lies in the knowledge of these five points.

孫
子

18. Hence the saying: **If you know the enemy and know yourself, you need not fear the result of a hundred battles. If you know yourself but not the enemy, for every victory gained you will also suffer a defeat.**

Li Ch'üan cites the case of Fu Chien, prince of Ch'in, who in AD 383 marched with a vast army against the Chin Emperor. When warned not to despise an enemy who could command the services of such men as Hsieh An and Huan Ch'ung, he boastfully replied: "I have the population of eight provinces at my back, infantry and horsemen to the number of one million; why, they could dam up the Yangtsze River itself by merely throwing their whips into the stream. What danger have I to fear?" Nevertheless, his forces were soon after disastrously routed at the Fei River, and he was obliged to beat a hasty retreat.

If you know neither the enemy nor yourself, you will succumb in every battle.

兵
法

[Chang Yü says words in this phrase] "have reference to attack and defense: knowing the enemy enables you to take the offensive, knowing yourself enables you to stand on the defensive." He adds: "Attack is the secret of defense; defense is the planning of an attack." It would be hard to find a better epitome of the root-principle of war.

TACTICAL DISPOSITIONS

[The wording of the title] is a very comprehensive and somewhat vague term. Literally, "form," "body," it comes to mean "appearance," "attitude" or "disposition"; and here it is best taken as something between, or perhaps combining, "tactics" and "disposition of troops." Ts'ao Kung explains it as "marching and countermarching on the part of the two armies with a view to discovering each other's condition." Tu Mu says: "It is through the dispositions of an army that its condition may be discovered. Conceal your dispositions, and your condition will remain secret, which leads to victory; show your dispositions, and your condition will become patent, which leads to defeat." Wang Hsi remarks that the good general can "secure success by modifying his tactics to meet those of the enemy."

孫
子

1. Sun Tzŭ said: The good fighters of old first put themselves beyond the possibility of defeat, and then waited for an opportunity of defeating the enemy.

2. To secure ourselves against defeat lies in our own hands, but the opportunity of defeating the enemy is provided by the enemy himself.

That is, of course, by a mistake on the enemy's part.

3. Thus the good fighter is able to secure himself against defeat,

"By concealing the disposition of his troops, covering up his tracks, and taking unremitting precautions" (Chang Yü).

but cannot make certain of defeating the enemy.

4. Hence the saying: One may *know* how to conquer without being able to *do* it.

5. Security against defeat implies defensive tactics; ability to defeat the enemy means taking the offensive.

   [For the beginning of this phrase] I retain the sense found in *sects.* 1–3, in spite of the fact that the commentators are all against me. The meaning they give, "He who cannot conquer takes the defensive," is plausible enough.

6. Standing on the defensive indicates insufficient strength; attacking, a superabundance of strength.

7. The general who is skilled in defense hides in the most secret recesses of the earth;

   Literally, "hides under the ninth earth," which is a metaphor indicating the utmost secrecy and concealment, so that the enemy may not know his whereabouts.

   he who is skilled in attack flashes forth from the topmost heights of heaven.

   Another metaphor, implying that he falls on his adversary like a thunderbolt, against which there is no time to prepare. This is the opinion of most of the commentators.

   Thus on the one hand we have ability to protect ourselves; on the other, a victory that is complete.

8. To see victory only when it is within the ken of the common herd is not the acme of excellence.

兵法

As Ts'ao Kung remarks, "the thing is to see the plant before it has germinated," to foresee the event before the action has begun. Li Ch'üan alludes to the story of Han Hsin who, when about to attack the vastly superior army of Chao, which was strongly entrenched in the city of Ch'êng-an, said to his officers: "Gentlemen, we are going to annihilate the enemy, and shall meet again at dinner." The officers hardly took his words seriously, and gave a very dubious assent. But Han Hsin had already worked out in his mind the details of a clever stratagem, whereby, as he foresaw, he was able to capture the city and inflict a crushing defeat on his adversary.

9. **Neither is it the acme of excellence if you fight and conquer and the whole Empire says, "Well done!"**

True excellence being, as Tu Mu says: "To plan secretly, to move surreptitiously, to foil the enemy's intentions and balk his schemes, so that at last the day may be won without shedding a drop of blood." Sun Tzŭ reserves his approbation for things that "the world's coarse thumb/And finger fail to plumb."

10. **To lift an autumn hair is no sign of great strength;**

["Autumn hair"] is explained as the fur of a hare, which is finest in autumn, when it begins to grow afresh. The phrase is a very common one in Chinese writers.

to see the sun and moon is no sign of sharp sight; to hear the noise of thunder is no sign of a quick ear.

Ho Shih gives as real instances of strength, sharp sight and quick hearing: Wu Huo, who could lift a tripod weighing 250 stone; Li Chu, who at a distance of a hundred paces could see objects no bigger than a mustard seed; and Shih K'uang, a blind musician who could hear the footsteps of a mosquito.

11. **What the ancients called a clever fighter is one who not only wins, but excels in winning with ease.**

Mei Yao-ch'ên says: "He who only sees the obvious, wins his battles with difficulty; he who looks below the surface of things, wins with ease."

12. **Hence his victories bring him neither reputation for wisdom nor credit for courage.**

Tu Mu explains this very well: "Inasmuch as his victories are gained over circumstances that have not come to light, the world at large knows nothing of them, and he wins no reputation for wisdom; inasmuch as the hostile state submits before there has been any bloodshed, he receives no credit for courage."

兵
法

13. **He wins his battles by making no mistakes.**

Ch'ên Hao says: "He plans no superfluous marches, he devises no futile attacks." The connection of ideas is thus explained by Chang Yü: "One who seeks to conquer by sheer strength, clever though he may be at winning pitched battles, is also liable on occasion to be vanquished; whereas he who can look into the future and discern conditions that are not yet manifest, will never make a blunder and therefore invariably win."

Making no mistakes is what establishes the certainty of victory, for it means conquering an enemy that is already defeated.

14. **Hence the skillful fighter puts himself into a position which makes defeat impossible, and does not miss the moment for defeating the enemy.**

A "counsel of perfection" as Tu Mu truly observes. [Position] need not be confined to the actual ground occupied by the troops. It includes all the arrangements and preparations which a wise general will make to increase the safety of his army.

15. **Thus it is that in war the victorious strategist only seeks battle after the victory has been won, whereas he who is destined to defeat first fights and afterwards looks for victory.**

Ho Shih thus expounds the paradox: "In warfare, first lay plans which will ensure victory, and then lead your army to battle; if you will not begin with stratagem but rely on brute strength alone, victory will no longer be assured."

16. **The consummate leader cultivates the moral law, and strictly adheres to method and discipline; thus it is in his power to control success.**

孫
子

17. In respect of military method, we have, firstly, Measurement; secondly, Estimation of quantity; thirdly, Calculation; fourthly, Balancing of chances; fifthly, Victory.

18. Measurement owes its existence to Earth; Estimation of quantity to Measurement; Calculation to Estimation of quantity; Balancing of chances to Calculation; and Victory to Balancing of chances.

It is not easy to distinguish the four terms very clearly. The first seems to be surveying and measurement of the ground, which enable us to form an estimate of the enemy's strength, and to make calculations based on the data thus obtained; we are thus led to [the fourth term], a general weighing-up, or comparison of the enemy's chances with our own; if the latter turn the scale, then victory ensues. The chief difficulty lies in [the third term], which some commentators take as a calculation of *numbers*, thereby making it nearly synonymous with [the second term]. Perhaps [the second term] should be thought of as a consideration of the enemy's general position or condition, while the third term is the estimate of his numerical strength.

兵
法

19. A victorious army opposed to a routed one, is as a pound's weight placed in the scale against a single grain.

Literally, "a victorious army is like an *i* (20 oz.) weighed against a *shu* (1/24 oz.); a routed army is a *shu* weighed against an *i*." The point is simply the enormous advantage which a disciplined force, flushed with victory, has over one demoralized by defeat.

20. The onrush of a conquering force is like the bursting of pent-up waters into a chasm a thousand fathoms deep. So much for tactical disposition.

ENERGY

1. **Sun Tzŭ said: The control of a large force is the same principle as the control of a few men: it is merely a question of dividing up their numbers.**

   That is, cutting up the army into regiments, companies, etc., with subordinate officers in command of each. Tu Mu reminds us of Han Hsin's famous reply to the first Han Emperor, who once said to him: "How large an army do you think I could lead?" "Not more than 100,000 men, your Majesty." "And you?" asked the Emperor. "Oh!" he answered, "the more the better."

2. **Fighting with a large army under your command is nowise different from fighting with a small one: it is merely a question of instituting signs and signals.**

3. **To ensure that your whole host may withstand the brunt of the enemy's attack and remain unshaken—this is effected by maneuvers direct and indirect.**

   We now come to one of the most interesting parts of Sun Tzŭ's treatise, the discussion of the [*chéng* and the *ch'i.*] As it is by no means easy to grasp the full significance of these two terms, or to render them consistently by good English equivalents; it may be as well to tabulate some of the commentators' remarks on the subject before proceeding further. Li Ch'üan: "Facing the enemy is *chéng*, making lateral diversion is *ch'i.*" Chia Lin: "In presence of the enemy, your troops should be arrayed in normal fashion, but in order to secure victory abnormal maneuvers must be employed." Mei Yao-ch'ên: "*Ch'i* is active, *chéng* is passive; passivity means waiting for an opportunity, activity beings the victory itself." Ho Shih: "We must cause the enemy to regard our straightforward attack as one that is secretly designed, and vice versa; thus *chéng* may also be *ch'i*, and *ch'i* may also be *chéng*." He instances the famous exploit of Han Hsin, who when marching ostensibly against Lin-chin (now Chao-i in Shensi), suddenly threw a large force across the Yellow River in wooden tubs, utterly disconcerting his opponent. (*Ch'ien Han Shu*, ch. 34.) Here, we are told, the march on Lin-chin was [*chéng*] and the surprise maneuver was [*ch'i*]. Chang Yü gives the following summary of opinions on the words: "Military writers do not agree with regard to the meaning of *ch'i* and *chéng*. Wei Liao Tzŭ (4th cent. BC) says: 'Direct warfare favors frontal attacks, indirect warfare attacks from the rear.' Ts'ao Kung says: 'Going straight out to join battle is a direct operation; appearing on the enemy's rear is an indirect maneuver.' Li Wei-kung (AD 6th and 7th cent.) says: 'In war, to march straight ahead is *chéng*; turning movements, on the other hand, are *ch'i*.'" These writers simply regard *chêng* as *chéng*, and *ch'i* as *ch'i*; they do not note that the two are mutually interchangeable and run into each other like the two sides of a circle (see *infra*, sect. 11). A comment on the T'ang Emperor T'ai Tsung goes to the root of the matter: "A *ch'i* maneuver may be *chéng*, if we make the enemy look upon it as *chéng*; then our real attack will be *ch'i*, and vice versa. The whole secret lies in confusing the enemy, so that he cannot fathom our real intent." To put it perhaps a little more clearly: any attack or other operation is *chéng*, on which the enemy has had his attention fixed; whereas that is ch'i, which takes him by surprise or comes from an unexpected quarter. If the enemy perceives a movement which is meant to be *ch'i*, it immediately becomes *chéng*.

4. **That the impact of your army may be like a grindstone dashed against an egg—this is effected by the science of weak points and strong.**

5. **In all fighting, the direct method may be used for joining battle, but indirect methods will be needed in order to secure victory.**

Chang Yü says: "Steadily develop indirect tactics, either by pounding the enemy's flanks or falling on his rear." A brilliant example of "indirect tactics" which decided the fortunes of a campaign was Lord

Roberts' night march round the Peiwar Kotal in the second Afghan war, [as recounted in *Forty-one Years in India*].

6. **Indirect tactics, efficiently applied, are inexhaustible as Heaven and Earth, unending as the flow of rivers and streams; like the sun and moon, they end but to begin anew; like the four seasons, they pass away to return once more.**

   Tu Yu and Chang Yü understand this of the permutations of [*ch'i*] and [*chéng*]. But at present Sun Tzŭ is not speaking of [*chéng*] at all, unless, indeed, we suppose with Chêng Yu-hsien that a clause relating to it has fallen out of the text. Of course, as has already been pointed out, the two are so inextricably interwoven in all military operations, that they cannot really be considered apart. Here we simply have an expression, in figurative language, of the almost infinite resource of a great leader.

孫
子

7. **There are not more than five musical notes, yet the combinations of these five give rise to more melodies than can ever be heard.**

8. **There are not more than five primary colors**

   Blue, yellow, red, white, and black.

   yet in combination they produce more hues than can ever been seen.

9. **There are not more than five cardinal tastes,**

   Sour, acrid, salt, sweet, bitter.

   yet combinations of them yield more flavors than can ever be tasted.

10. In battle, there are not more than two methods of attack—
the direct and the indirect; yet these two in combination give
rise to an endless series of maneuvers.

11. The direct and the indirect lead on to each other in turn. It is
like moving in a circle—you never come to an end. Who can
exhaust the possibilities of their combination?

12. The onset of troops is like the rush of a torrent which will
even roll stones along in its course.

13. The quality of decision is like the well-timed swoop of a
falcon which enables it to strike and destroy its victim.

> [The wording here defies the best efforts of the translator. Tu Mu
> defines this word as] "the measurement or estimation of distance."
> But this meaning does not quite fit the illustrative simile in sect. 15.
> Applying this definition to the falcon, it seems to me to denote that
> instinct of *self-restraint* which keeps the bird from swooping on its
> quarry until the right moment, together with the power of judging
> when the right moment has arrived. The analogous quality in sol-
> diers is the highly important one of being able to reserve their fire
> until the very instant at which it will be most effective. When the
> "Victory" went into action at Trafalgar at hardly more than drifting
> pace, she was for several minutes exposed to a storm of shot and shell
> before replying with a single gun. Nelson coolly waited until he was
> within close range, when the broadside he brought to bear worked
> fearful havoc on the enemy's nearest ships.

兵
法

14. Therefore the good fighter will be terrible in his onset, and
prompt in his decision.

15. **Energy may be likened to the bending of a crossbow; decision, to the releasing of a trigger.**

"Energy" seems to be the best equivalent here for [the wording], because the comparison implies that the force is potential, being stored up in the bent cross-bow until released by the finger on the trigger. None of the commentators seem to grasp the real point of the simile.

16. **Amid the turmoil and tumult of battle, there may be seeming disorder and yet no real disorder at all; amid confusion and chaos, your array may be without head or tail, yet it will be proof against defeat.**

["Without head or tail" derives literally from] "formation circular," which is explained by Li Ch'üan as "without back or front." Mei Yao-ch'ên says: "The subdivisions of the army having been previously fixed, and the various signals agreed upon, the separating and joining, the dispersing and collecting which will take place in the course of a battle, may give the appearance of disorder when no real disorder is possible. Your formation may be without head or tail, your dispositions all topsy-turvy, and yet a rout of your forces quite out of the question."

17. **Simulated disorder postulates perfect discipline, simulated fear postulates courage; simulated weakness postulates strength.**

In order to make the translation intelligible, it is necessary to tone down the sharply paradoxical form of the original. Ts'ao Kung throws out a hint of the meaning in his brief note: "These things all serve to destroy formation and conceal one's condition." But Tu Mu is the first to put it quite plainly: "If you wish to feign confusion in order to lure the enemy on, you must first have perfect discipline; if you wish to display timidity in order to entrap the enemy, you must have extreme courage; if you wish to parade your weakness in order to make the enemy over-confident, you must have exceeding strength."

18. Hiding order beneath the cloak of disorder is simply a question of subdivision;

See *supra*, sect. 1.

concealing courage under a show of timidity presupposes a fund of latent energy; masking strength with weakness is to be effected by tactical dispositions.

Chang Yü relates the following anecdote of Kao Tsu, the first Han Emperor: "Wishing to crush the Hsiung-nu, he sent out spies to report on their condition. But the Hsiung-nu, forewarned, carefully concealed all their able-bodied men and well-fed horses, and only allowed infirm soldiers and emaciated cattle to be seen. The result was that spies one and all recommended the Emperor to deliver his attack. Lu Ching alone opposed them, saying: 'When two countries go to war, they are naturally inclined to make an ostentatious display of their strength. Yet our spies have seen nothing but old age and infirmity. This is surely some *ruse* on the part of the enemy, and it would be unwise for us to attack.' The Emperor, however, disregarding this advice, fell into the trap and found himself surrounded at Po-têng."

19. Thus one who is skillful at keeping the enemy on the move maintains deceitful appearances, according to which the enemy will act.

Ts'ao Kung's note is "Make a display of weakness and want." Tu Mu says: "If our force happens to be superior to the enemy's, weakness may be simulated in order to lure him on; but if inferior, he must be led to believe that we are strong, in order that he may keep off. In fact, all the enemy's movements should be determined by the signs that we choose

121

to give him." Note the following anecdote of Sun Pin, a descendent of Sun Wu: In 341 BC, the Ch'i State being at war with Wei, sent T'ien Chi and Sun Pin against the general P'ang Chüan, who happened to be a deadly personal enemy of the latter. Sun Pin said: "The Ch'i State has a reputation for cowardice, and therefore our adversary despises us. Let us turn this circumstance to account." Accordingly, when the army had crossed the border into Wei territory, he gave orders to show 100,000 fires on the first night, 50,000 on the next, and the night after only 20,000. P'ang Chüan pursued them hotly, saying to himself: "I knew these men of Ch'i were cowards: their numbers have already fallen away by more than half." In his retreat, Sun Pin came to a narrow defile, with he calculated that his pursuers would reach after dark. Here he had a tree stripped of its bark, and inscribed upon it the words: "Under this tree shall P'ang Chüan die." Then, as night began to fall, he placed a strong body of archers in ambush near by, with orders to shoot directly they saw a light. Later on, P'ang Chuan arrived at the spot, and noticing the tree, struck a light in order to read what was written on it. His body was immediately riddled by a volley of arrows, and his whole army thrown into confusion. (The above is Tu Mu's version of the story; the *shih chi*, less dramatically but probably with more historical truth, makes P'ang Chuan cut his own throat with an exclamation of despair, after the rout of his army.)

He sacrifices something, that the enemy may snatch at it.

20. **By holding out baits, he keeps him on the march; then with a body of picked men he lies in wait for him.**

[With an emendation suggested by Li Ching] the meaning then would be, "He lies in wait with the main body of his troops."

21. **The clever combatant looks to the effect of combined energy, and does not require too much from individuals.**

Tu Mu says: "He first of all considers the power of his army in the bulk; afterwards he takes individual talent into account, and uses each man according to his capabilities. He does not demand perfection from the untalented."

Hence his ability to pick out the right men and utilize combined energy.

22. **When he utilizes combined energy, his fighting men become as it were like unto rolling logs or stones. For it is the nature of a log or stone to remain motionless on level ground, and to move when on a slope; if four-cornered, to come to a standstill, but if round-shaped, to go rolling down.**

Ts'au Kung calls this "the use of natural or inherent power."

23. **Thus the energy developed by good fighting men is as the momentum of a round stone rolled down a mountain thousands of feet in height. So much on the subject of energy.**

The chief lesson of this chapter, in Tu Mu's opinion, is the paramount importance in war of rapid evolutions and sudden rushes. "Great results," he adds, "can thus be achieved with small forces."

WEAK POINTS
AND STRONG

Chang Yü attempts to explain the sequence of chapters as follows: "Chapter IV, on Tactical Dispositions, treated of the offensive and the defensive; chapter V, on Energy, dealt with direct and indirect methods. The good general acquaints himself first with the theory of attack and defense, and then turns his attention to direct and indirect methods. He studies the art of varying and combining these two methods before proceeding to the subject of weak and strong points. For the use of direct or indirect methods arises out of attack and defense, and the perception of weak and strong points depends again on the above methods. Hence the present chapter comes immediately after the chapter on Energy."

1. Sun Tzŭ said: Whoever is first in the field and awaits the coming of the enemy, will be fresh for the fight; whoever is second in the field and has to hasten to battle will arrive exhausted.

2. Therefore the clever combatant imposes his will on the enemy, but does not allow the enemy's will to be imposed on him.

   One mark of a great soldier is that he fights on his own terms or fights not at all.

3. By holding out advantages to him, he can cause the enemy to approach of his own accord; or, by inflicting damage, he can make it impossible for the enemy to draw near.

   In the first case, he will entice him with a bait; in the second, he will strike at some important point which the enemy will have to defend.

4. **If the enemy is taking his ease, he can harass him; if well supplied with food, he can starve him out; if quietly encamped, he can force him to move.**

5. **Appear at points which the enemy must hasten to defend; march swiftly to places where you are not expected.**

[T'sao Kung and Ho Shih read this as]: "Appear at points to which the enemy cannot hasten."

6. **An army may march great distances without distress, if it marches through country where the enemy is not.**

Ts'ao Kung sums up very well: "Emerge from the void (*q.d.*, like "a bolt from the blue"), strike at vulnerable points, shun places that are defended, attack in unexpected quarters."

7. **You can be sure of succeeding in your attacks if you only attack places which are undefended.**

Wang Hsi rightly explains [undefended places] as "weak points; that is to say, where the general is lacking in capacity, or the soldiers in spirit; where the walls are not strong enough, or the precautions not strict enough; where relief comes too late, or provisions are too scanty, or the defenders are variance amongst themselves."

You can ensure the safety of your defense if you only hold positions that cannot be attacked.

I.e., where there are none of the weak points mentioned above. There is rather a nice point involved in the interpretation of this later clause. Tu Mu, Ch'ên Hao, and Mei Yao-ch'ên assume the meaning to be: "In order to make your defense quite safe, you must defend even those places that are not likely to be attacked"; and Tu Mu adds: "How much more, then, those that will be attacked." Taken thus, however, the clause balances less well with the preceding—always a consideration in the highly antithetical style which is natural to the Chinese. Chang Yü, therefore, seems to come nearer the mark in saying: "He who is skilled in attack flashes forth from the topmost heights of heaven, making it impossible for the enemy to guard against him. This being so, the places that I shall attack are precisely those that the enemy cannot defend. . . . He who is skilled in defense hides in the most secret recesses of the earth, making it impossible for the enemy to estimate his whereabouts. This being so, the places that I shall hold are precisely those that the enemy cannot attack."

8. **Hence that general is skillful in attack whose opponent does not know what to defend; and he is skillful in defense whose opponent does not know what to attack.**

An aphorism which puts the whole art of war in a nutshell.

9. **O divine art of subtlety and secrecy! Through you we learn to be invisible, through you inaudible;**

Literally, "without form or sound," but it is said of course with reference to the enemy.

and hence we can hold the enemy's fate in our hands.

10. **You may advance and be absolutely irresistible, if you make for the enemy's weak points; you may retire and be safe from pursuit if your movements are more rapid than those of the enemy.**

11. **If we wish to fight, the enemy can be forced to an engagement even though he be sheltered behind a high rampart and a deep ditch. All we need do is attack some other place that he will be obliged to relieve.**

Tu Mu says: "If the enemy is the invading party, we can cut his line of communications and occupy the roads by which he will have to return; if we are the invaders, we may direct our attack against the sovereign himself." It is clear that Sun Tzŭ, unlike certain generals in the late Boer war, was no believer in frontal attacks.

12. **If we do not wish to fight, we can prevent the enemy from engaging us even though the lines of our encampment be merely traced out on the ground. All we need do is to throw something odd and unaccountable in his way.**

This extremely concise expression is intelligibly paraphrased by Chia Lin: "even though we have constructed neither wall nor ditch." . . . Li Ch'üan says: "we puzzle him by strange and unusual dispositions"; and Tu Mu finally clinches the meaning by three illustrative anecdotes—one of Chu-ko Liang, who when occupying Yang-p'ing and about to be attacked by Ssŭ-ma I, suddenly struck his colors, stopped the beating of the drums, and flung open the city gates, showing only a few men engaged in sweeping and sprinkling the ground. This unexpected proceeding had the intended effect; for Ssŭ-ma I, suspecting an ambush, actually drew off his army and retreated. What Sun Tzŭ is advocating here, therefore, is nothing more nor less than the timely use of "bluff."

兵
法

13. **By discovering the enemy's dispositions and remaining invisible ourselves, we can keep our forces concentrated, while the enemy's must be divided.**

The conclusion is perhaps not very obvious, but Chang Yü (after Mei Yao-ch'ên) rightly explains it thus: "If the enemy's dispositions are visible, we can make for him in one body; whereas, our own dispositions being kept secret, the enemy will be obliged to divide his forces in order to guard against attack from every quarter."

14. We can form a single united body, while the enemy must split up into fractions. Hence there will be a whole pitted against separate parts of a whole, which means that we shall be many to the enemy's few.

15. And if we are able thus to attack an inferior force with a superior one, our opponents will be in dire straits.

16. The spot where we intend to fight must not be made known; for then the enemy will have to prepare against a possible attack at several different points;

    Sheridan once explained the reason of General Grant's victories by saying that "while his opponents were kept fully employed wondering what he was going to do, *he* was thinking most of what he was going to do himself."

    and his forces being thus distributed in many directions, the numbers we shall have to face at any given point will be proportionately few.

17. For should the enemy strengthen his van, he will weaken his rear; should he strengthen his rear, he will weaken his van; should he strengthen his left, he will weaken his right; should he strengthen his right, he will weaken his left. If he sends reinforcements everywhere, he will everywhere be weak.

    In Frederick the Great's *Instructions to His Generals* we read: "A defensive war is apt to betray us into too frequent detachment. Those generals who have had but little experience attempt to protect every point, while those who are better acquainted with their profession, having only the capital object in view, guard against a decisive blow, and acquiesce in small misfortunes to avoid greater."

18. **Numerical weakness comes from having to prepare against possible attacks; numerical strength, from compelling our adversary to make these preparations against us.**

    The highest generalship, in Col. Henderson's words, is "to compel the enemy to disperse his army, and then to concentrate superior force against each fraction in turn."

19. **Knowing the place and the time of the coming battle, we may concentrate from the greatest distances in order to fight.**

    What Sun Tzŭ evidently has in mind is that nice calculation of distances and that masterly employment of strategy which enable a general to divide his army for the purpose of a long and rapid march, and afterwards to effect a junction at precisely the right spot and the right hour in order to confront the enemy in overwhelming strength. Among many such successful junctions which military history records, one of the most dramatic and decisive was the appearance of Blücher just at the critical moment on the field of Waterloo.

兵
法

20. **But if neither time nor place be known, then the left wing will be impotent to succor the right, the right equally impotent to succor the left, the van unable to relieve the rear, or the rear to support the van. How much more so if the furthest portions of the army are anything under a hundred *li* apart, and even the nearest are separated by several *li*!**

The Chinese of this last sentence is a little lacking in precision, but the mental picture we are required to draw is probably that of an army advancing towards a given rendezvous in separate columns, each of which has orders to be there on a fixed date. If the general allows the various detachments to proceed at haphazard, without precise instructions as to the time and place of meeting, the enemy will be able to annihilate the army in detail. Chang Yü's note may be worth quoting here: "If we do not know the place where our opponents mean to concentrate or the day on which they will join battle, our unity will be forfeited through our preparations for defense, and the positions we hold will be insecure. Suddenly happening upon a powerful foe, we shall be brought to battle in a flurried condition, and no mutual support will be possible between wings, vanguard or rear, especially if there is any great distance between the foremost and hindmost divisions of the army."

21. **Though according to my estimate the soldiers of Yüeh exceed our own in number, that shall advantage them nothing in the matter of victory. I say then that victory can be achieved.**

Alas for these brave words! The long feud between the two states ended in 473 BC with the total defeat of Wu by Kou Chien and its incorporation in Yüeh. This was doubtless long after Sun Tzŭ's death. With his present assertion compare ch. IV, sect. 4. . . . Chang Yü is the only one to point out the seeming discrepancy, which he thus goes on to explain: "In the chapter on Tactical Dispositions it is said, 'One may *know* how to conquer without being able to *do* it,' whereas here we have the statement that 'victory' can be 'achieved.' The explanation is, that in the former chapter, where the offensive and defensive are under discussion, it is said that if the enemy is fully prepared, one cannot make certain of beating him. But the present passage refers particularly to the soldiers of Yüeh who, according to Sun Tzŭ's calculations, will be kept in ignorance of the time and place of the impending struggle. That is why he says here that victory can be achieved."

22. **Though the enemy be stronger in numbers, we may prevent him from fighting. Scheme so as to discover his plans and the likelihood of their success.**

[An alternative reading offered by Chia Lin is]: "Know beforehand all plans conducive to our success and to the enemy's failure."

23. **Rouse him, and learn the principle of his activity or inactivity.**

. . . Chang Yü tells us that by noting the joy or anger shown by the enemy on being thus disturbed, we shall be able to conclude whether his policy is to lie low or the reverse. He instances the action of Cho-ku Liang, who sent the scornful present of a woman's head-dress to Ssŭ-ma I, in order to goad him out of his Fabian tactics.

兵
法

Force him to reveal himself, so as to find out his vulnerable spots.

24. **Carefully compare the opposing army with your own, so that you may know where strength is superabundant and where it is deficient.**

Cf. ch. IV, sect. 6.

25. **In making tactical dispositions, the highest pitch you can attain is to conceal them;**

The piquancy of the paradox evaporates in translation. [Conceal] is perhaps not so much actual invisibility (see *supra* sect. 9) as "showing no sign" of what you mean to do, of the plans that are formed in your brain.

conceal your dispositions, and you will be safe from the prying of the subtlest spies, from the machinations of the wisest brains.

[Tu Mu explains]: "Though the enemy may have clever and capable officers, they will not be able to lay any plans against us."

26. How victory may be produced for them out of the enemy's own tactics—that is what the multitude cannot comprehend.

27. All men can see the tactics whereby I conquer, but what none can see is the strategy out of which victory is evolved.

I.e., everybody can see superficially how a battle is won; what they cannot see is the long series of plans and combinations which has preceded the battle.

28. Do not repeat the tactics which have gained you one victory, but let your methods be regulated by the infinite variety of circumstances.

As Wang Hsi sagely remarks: "There is but one root-principle underlying victory, but the tactics which lead up to it are infinite in number." With this compare Col. Henderson: "The rules of strategy are few and simple. They may be learned in a week. They may be taught by familiar illustrations or a dozen diagrams. But such knowledge will no more teach a man to lead an army like Napoleon than a knowledge of grammar will teach him to write like Gibbon."

29. Military tactics are like unto water; for water in its natural course runs away from high places and hastens downwards.

30. So in war, the way is to avoid what is strong and to strike at what is weak.

Like water, taking the line of least resistance.

孫
子

31. Water shapes its course according to the nature of the ground over which it flows; the soldier works out his victory in relation to the foe whom he is facing.

32. Therefore, just as water retains no constant shape, so in warfare there are no constant conditions.

33. He who can modify his tactics in relation to his opponent and thereby succeed in winning, may be called a heaven-born captain.

34. The five elements

Water, fire, wood, metal, earth.

are not always predominant;

That is, as Wang Hsi says: "they predominate alternately."

the four seasons make way for each other in turn.

Literally, "have no invariable seat."

There are short days and long; the moon has its periods of waning and waxing.

Cf. ch. V, sect. 6. The purport of the passage is simply to illustrate the want of fixity in war by the changes constantly taking place in Nature. The comparison is not very happy, however, because the regularity of the phenomena which Sun Tzŭ mentions is by no means paralleled in war.

MANEUVERING

The commentators, as well as the subsequent text, make it clear that [maneuvering] is the real meaning of [the chapter title]. Thus, Li Ch'üan [interprets it to mean] "marching rapidly to seize an advantage"; Wang Hsi says: "'Striving' means striving for an advantage; this being obtained, victory will follow"; and Chang Yü: "The two armies face to face, and each striving to obtain a tactical advantage over the other." According to the latter commentator, then, the situation is analogous to that of two wrestlers maneuvering for a "hold," before coming to actual grips.

孫子

1. Sun Tzǔ said: In war, the general receives his commands from the sovereign.

2. Having collected an army and concentrated his forces, he must blend and harmonize the different elements thereof before pitching his camp.

   Chang Yü [says this section] may also be taken to mean "the establishment of harmony and confidence between the higher and lower ranks before venturing into the field"; and he quotes a saying of Wu Tzǔ (ch. 1 *ad init.*): "Without harmony in the State, no military expedition can be undertaken; without harmony in the army, no battle array can be formed." In an historical romance Sun Tzǔ is represented as saying to Wu Yüan: "As a general rule, those who are waging war should get rid of all the domestic troubles before proceeding to attack the external foe."

3. After that, comes tactical maneuvering, than which there is nothing more difficult.

I have departed slightly from the traditional interpretation of Ts'ao Kung, who says: "From the time of receiving the sovereign's instructions until our encampment over against the enemy, the tactics to be pursued are most difficult." It seems to me that the tactics or maneuvers can hardly be said to begin until the army has sallied forth and encamped, and Ch'iên Hao's note gives color to this view: "For levying, concentrating, harmonizing and entrenching an army, there are plenty of old rules which will serve. The real difficulty comes when we engage in tactical operations." Tu Yu also observes that "the great difficulty is to be beforehand with the enemy in seizing favorable position."

The difficulty of tactical maneuvering consists in turning the devious into the direct, and misfortune into gain.

[This sentence contains] one of those highly condensed and somewhat enigmatical expressions of which Sun Tzŭ is so fond. This is how it is explained by Ts'ao Kung: "Make it appear that you are a long way off, then cover the distance rapidly and arrive on the scene before your opponent." Tu Mu says: "Hoodwink the enemy, so that he may be remiss and leisurely while you are dashing along with utmost speed." Ho Shih gives a slightly different turn: "Although you may have difficult ground to traverse and natural obstacles to encounter this is a drawback which can be turned into actual advantage by celerity of movement." Signal examples of this saying are afforded by the two famous passages across the Alps—that of Hannibal, which laid Italy at his mercy, and that of Napoleon two thousand years later, which resulted in the great victory of Marengo.

兵
法

4. **Thus, to take a long and circuitous route, after enticing the enemy out of the way, and though starting after him, to contrive to reach the goal before him, shows knowledge of the artifice of *deviation*.**

Chia Lin understands [the phrasing] as the enemy's line of march, thus: "If our adversary's course is really a short one, and we can manage to divert him from it either by simulating weakness or holding out some small advantage, we shall be able to beat him in the race for good positions." This is quite a defensible view, though not adopted by any other commentator. . . . Tu Mu cites the famous march of Chao Shê in 270 BC to relieve the town of O-yü, which was closely invested by a Ch'in army. . . . The King of Chao first consulted Lien P'o on the advisability of attempting a relief, but the latter thought the distance too great, and the intervening country too rugged and difficult. His Majesty then turned to Chao Shê, who fully admitted the hazardous nature of the march, but finally said: "We shall be like two rats fighting in a hole—and the pluckier one will win!" So he left the capital with his army, but had only gone a distance of 30 *li* when he stopped and began throwing up entrenchments. For 28 days he continued strengthening his fortifications, and took care that spies should carry the intelligence to the enemy. The Ch'in general was overjoyed, and attributed his adversary's tardiness to the fact that the beleaguered city was in the Han State, and thus not actually part of Chao territory. But the spies had no sooner departed than Chao Shê began a forced march lasting for two days and one night, and arrive on the scene of action with such astonishing rapidity that he was able to occupy a commanding position on the "North hill" before the enemy had got wind of his movements. A crushing defeat followed for the Ch'in forces, who were obliged to raise the siege of O-yü in all haste and retreat across the border.

5. **Maneuvering with an army is advantageous; with an undisciplined multitude, most dangerous.**

I here adopt the reading of the *T'ung Tien*, Chêng Yu-hsien and the T'u Shu . . . The commentators take it to mean that maneuvers may be profitable, or they may be dangerous: it all depends on the ability of the general.

6. **If you set a fully equipped army in march in order to snatch an advantage, the chances are that you will be too late. On the other hand, to detach a flying column for the purpose involves the sacrifice of its baggage and stores.**

    [Some of the text is unintelligible to the Chinese commentators.] On the whole, it is clear that Sun Tzǔ does not approve of a lengthy march being undertaken without supplies. *Cf. infra* sect. 11.

7. **Thus, if you order your men to roll up their buff-coats, and make forced marches without halting day or night, covering double the usual distance at a stretch,**

    The ordinary day's march, according to Tu Mu, was 30 *li*; but on one occasion, when pursuing Liu Pei, Ts'ao Ts'ao is said to have covered the incredible distance of 300 *li* within twenty-four hours.

    doing a hundred *li* in order to wrest an advantage, the leaders of all your three divisions will fall into the hands of the enemy.

    兵
    法

8. **The stronger men will be in front, the jaded ones will fall behind, and on this plan only one-tenth of your army will reach its destination.**

    The moral is, as Ts'ao Kung and others point out: Don't march a hundred *li* to gain a tactical advantage, either with or without impedimenta. Maneuvers of this description should be confined to short distances. Stonewall Jackson said: "The hardships of forced marches are often more painful than the dangers of battle." He did not often call upon his troops for extraordinary exertions. It was only when he intended a surprise, or when a rapid retreat was imperative, that he sacrificed everything for speed.

9. **If you march fifty *li* in order to outmaneuver the enemy, you will lose the leader of your first division, and only half your force will reach the goal.**

[L]iterally, "the leader of the first division will be *torn away*."

10. If you march thirty *li* with the same object, two-thirds of your army will arrive.

   In the *T'ung Tien* is added: "From this we may know the difficulty of maneuvering."

11. We may take it then that an army without its baggage-train is lost; without provisions it is lost; without bases of supply it is lost.

   [I think Sun Tzŭ meant "stores accumulated in depots." But Tu Yu says "fodder and the like," Chang Yü says "Goods in general," and Wang Hsi says "fuel, salt, foodstuffs, etc."]

孫
子

12. We cannot enter into alliances until we are acquainted with the designs of our neighbors.

13. We are not fit to lead an army on the march unless we are familiar with the face of the country—its mountains and forests, its pitfalls and precipices, its marshes and swamps.

14. We shall be unable to turn natural advantage to account unless we make use of local guides.

   Sects. 12–14 are repeated in ch. XI, sect. 52.

15. In war, practice dissimulation, and you will succeed. Move only if there is a real advantage to be gained.

16. Whether to concentrate or to divide your troops, must be decided by circumstances.

17. **Let your rapidity be that of the wind,**

The simile is doubly appropriate, because the wind is not only swift but, as Mei Yao-ch'ên points out, "invisible and leaves no tracks."

your compactness that of the forest.

Mêng Shih [says]: "When slowly marching, order and ranks must be preserved"—so as to guard against surprise attacks. But natural forests do not grow in rows, whereas they do generally possess the quality of density or compactness.

18. **In raiding and plundering be like fire,**

Cf. Shih Ching IV. 3. iv. 6: "Fierce as a blazing fire which no man can check."

in immovability like a mountain.

That is, when holding a position from which the enemy is trying to dislodge you, or perhaps, as Tu Yu says, when he is trying to entice you into a trap.

19. **Let your plans be dark and impenetrable as night, and when you move, fall like a thunderbolt.**

Tu Yu quotes a saying of T'ai Kung which has passed into a proverb: "You cannot shut your ears to the thunder or your eyes to the lighting—so rapid are they." Likewise, an attack should be made so quickly that it cannot be parried.

20. **When you plunder a countryside, let the spoil be divided amongst your men;**

Sun Tzŭ wishes to lessen the abuses of indiscriminate plundering by insisting that all booty shall be thrown into a common stock, which may afterwards be fairly divided amongst all.

when you capture new territory, cut it up into allotments for the benefit of the soldiery.

Ch'ên Hao [says] "quarter your soldiers on the land, and let them sow and plant it." It is by acting on this principle, and harvesting the lands they invaded, that the Chinese have succeeded in carrying out some of their most memorable and triumphant expeditions, such as that of Pan Ch'ao who penetrated to the Caspian, and in more recent years, those of Fu-k'ang-an and Tso Tsung-T'ang.

21. **Ponder and deliberate before you make a move.**

Chang Yü quotes [Wei Liao Tzŭ] as saying that we must not break camp until we have gained the resisting power of the enemy and the cleverness of the opposing general. *Cf.* the "seven comparisons" in ch. I, sect. 13.

22. **He will conquer who has learned the artifice of deviation.**

See *supra*, sects. 3, 4.

Such is the art of maneuvering.

With these words, the chapter would naturally come to an end. But there now follows a long appendix in the shape of an extract from an earlier book on War, now lost, but apparently extant at the time when Sun Tzŭ wrote. The style of this fragment is not noticeable different from that of Sun Tzŭ himself, but no commentator raises a doubt as to its genuineness.

23. **The Book of Army Management says:**

It is perhaps significant that none of the earlier commentators give us any information about this work. Mei Yao-ch'ên calls it "an ancient military classic," and Wang Hsi, "an old book on war." Considering the enormous amount of fighting that had gone on for centuries before

孫
子

Sun Tzŭ's time between the various kingdoms and principalities of China, it is not in itself improbable that a collection of military maxims should have been made and written down at some earlier period.

On the field of battle,

Implied, though not actually in the Chinese.

the spoken word does not carry far enough: hence the institution of gongs and drums. Nor can ordinary objects be seen clearly enough: hence the institution of banners and flags.

24. **Gongs and drums, banners and flags, are means whereby the ears and eyes of the host may be focused on one particular point.**

Chang Yü says: "If sight and hearing converge simultaneously on the same object, the evolutions of as many as a million soldiers will be like those of a single man!"

25. **The host thus forming a single united body, is it impossible either for the brave to advance alone, or for the cowardly to retreat alone.**

Chuang Yu quotes a saying: "Equally guilty are those who advance against orders and those who retreat against orders." Tu Mu tells a story in this connection of Wu Ch'i, when he was fighting against the Ch'in State. Before the battle had begun, one of his soldiers, a man of matchless daring, sallied forth by himself, captured two heads from the enemy, and returned to camp. Wu Ch'i had the man instantly executed, whereupon an officer ventured to remonstrate, saying: "This man was a good soldier, and ought not to have been beheaded." Wu Ch'i replied: "I fully believe he was a good soldier, but I had him beheaded because he acted without orders."

兵
法

This is the art of handling large masses of men.

26. **In night-fighting, then, make much use of signal-fires and drums, and in fighting by day, of flags and banners, as a means of influencing the ears and eyes of your army.**

Ch'ên Hao alludes to Li Kuang-pi's night ride to Ho-yang at the head of 500 mounted men; they made such an imposing display with torches, that though the rebel leader Shih Ssŭ-ming had a large army, he did not dare to dispute their passage.

27. **A whole army may be robbed of its spirit;**

"In war," says Chang Yü, "if a spirit of anger can be made to pervade all ranks of an army at one and the same time, its onset will be irresistible. Now the spirit of the enemy's soldiers will be keenest when they have newly arrived on the scene, and it is therefore our cue not to fight at once, but to wait until their ardor and enthusiasm have worn off, and then strike. It is in this way that they may be robbed of their keen spirit." Li Ch'üan and others tell an anecdote . . . of Ts'ao Kuei, a protégé of Duke Chuang of Lu. The latter State was attacked by Ch'i, and the duke was about to join battle at Ch'ang-cho, after the first roll of the enemy's drums, when Ts'ao said: "Not just yet." Only after their drums had beaten for the third time, did he give the word for attack. Then they fought, and the men of Ch'i were utterly defeated. Questioned afterwards by the Duke as to the meaning of his delay, Ts'ao Kuei replied: "In battle, a courageous spirit is everything. Now the first roll of the drum tends to create this spirit, but with the second it is already on the wane, and after the third it is gone altogether. I attacked when their spirit was gone and ours was at its height. Hence our victory." Wu Tzŭ . . . puts "spirit" first among the "four important influences" in war, and continues: "The value of a whole army—a mighty host of a million men—is dependent on one man alone: such is the influence of spirit!"

a commander-in-chief may be robbed of his presence of mind.

Chang Yü says: "Presence of mind is the general's most important asset. It is the quality which enables him to discipline disorder and to inspire courage into the panic-stricken." The great general Li Ching (AD 571–649) has a saying: "Attacking does not merely consist in assaulting walled cities or striking at an army in battle array; it must include the art of assailing the enemy's mental equilibrium."

28. **Now a soldier's spirit is keenest in the morning; by noonday it has begun to flag; and in the evening, his mind is bent only on returning to camp.**

Always provided, I suppose, that he has had breakfast. At the battle of Trebia, the Romans were foolishly allowed to fight fasting, whereas Hannibal's men had breakfasted at their leisure.

29. **A clever general, therefore, avoids an army when its spirit is keen, but attacks it when it is sluggish and inclined to return. This is the art of studying moods.**

兵
法

30. **Disciplined and calm, to await the appearance of disorder and hubbub amongst the enemy:—this is the art of retaining self-possession.**

31. **To be near the goal while the enemy is still far from it, to wait at ease while the enemy is toiling and struggling, to be well-fed while the enemy is famished:—this is the art of husbanding one's strength.**

32. **To refrain from intercepting an enemy whose banners are in perfect order, to refrain from attacking an army drawn up in calm and confident array:—this is the art of studying circumstances.**

33. **It is a military axiom not to advance uphill against the enemy, nor to oppose him when he comes downhill.**

34. **Do not pursue an enemy who simulates flight; do not attack soldiers whose temper is keen.**

35. **Do not swallow bait offered by the enemy.**

    Li Ch'üan and Tu Mu, with extraordinary inability to see a metaphor, take these words quite literally of food and drink that have been poisoned by the enemy. Ch'ên Hao and Chang Yü carefully point out that the saying has a wider application.

    Do not interfere with an army that is returning home.

    The commentators explain this rather singular piece of advice by saying that a man whose heart is set on returning home will fight to the death against any attempt to bar his way, and is therefore too dangerous an opponent to be tackled. Chang Yü quotes the words of Han Hsin: "Invincible is the soldier who hath his desire and returneth homewards." A marvelous tale is told of Ts'ao Ts'ao's courage and resource in the *San Kuo Chi*: In AD 198, he was besieging Chang Hsiu in Jang, when Liu Piao sent reinforcements with a view to cutting off Ts'ao's retreat. The latter was obliged to draw off his troops, only to find himself hemmed in between two enemies, who were guarding each outlet of a narrow pass in which he had engaged himself. In this desperate plight Ts'ao waited until nightfall, when he bored a tunnel into the mountain side and laid an ambush in it. As soon as the whole army had passed by, the hidden troops fell on his rear, while Ts'ao himself turned and met his pursuers in front, so that they were thrown into confusion and annihilated. Ts'ao Ts'ao said afterwards: "The brigands tried to check my army in its retreat and brought me to battle in a desperate position: hence I knew how to overcome them."

36. **When you surround an army, leave an outlet free.**

This does not mean that the enemy is to be allowed to escape. The object, as Tu Mu puts it, is "to make him believe that there is a road to safety, and thus prevent his fighting with the courage of despair." Tu Mu adds pleasantly: "After that, you may crush him."

Do not press a desperate foe too hard.

Ch'ên Hao quotes the saying: "Birds and beasts when brought to bay will use their claws and teeth." Chang Yü says: "If your adversary has burned his boats and destroyed his cooking-pots, and is ready to stake all on the issue of a battle, he must not be pushed to extremities." . . . Ho Shih illustrates the meaning by a story taken from the life of Yen-ch'ing. That general, together with his colleague Tu Chung-wei was surrounded by a vastly superior army of Khitans in the year AD 945. The country was bare and desert-like, and the little Chinese force was soon in dire straits for want of water. The wells they bored ran dry, and the men were reduced to squeezing lumps of mud and sucking out the moisture. Their ranks thinned rapidly, until at last Fu Yen-ch'ing exclaimed: "We are desperate men. Far better to die for our country than to go with fettered hands into captivity!" A strong gale happened to be blowing from the northeast and darkening the air with dense clouds of sandy dust. Tu Chung-wei was for waiting until this had abated before deciding on a final attack; but luckily another officer, Li Shou-chêng by name, was quicker to see an opportunity, and said: "They are many and we are few, but in the midst of this sandstorm our numbers will not be discernible; victory will go to the strenuous fighter, and the wind will be our best ally." Accordingly, Fu Yen-ch'ing made a sudden and wholly unexpected onslaught with his cavalry, routed the barbarians and succeeded in breaking through to safety.

兵
法

37. **Such is the art of warfare.**

VARIATION IN TACTICS

The heading means literally "The Nine Variations," but as Sun Tzŭ does not appear to enumerate these, and as, indeed, he has already told us (ch. V, sects. 6–11) that such deflections from the ordinary course are practically innumerable, we have little option but to follow Wang Hsi, who says that "Nine" stands for an indefinitely large number. "All it means is that in warfare we ought to vary our tactics to the utmost degree. . . . I do not know what Ts'ao Kung makes these Nine Variations out to be, but it has been suggested that they are connected with the Nine Situations." This is the view adopted by Chang Yü. The only other alternative is to suppose that something has been lost—a supposition to which the unusual shortness of the chapter lends some weight.

1. Sun Tzŭ said: In war, the general receives his commands from the sovereign, collects his army and concentrates his forces.

   Repeated from ch. VII, sect. 1, where it is certainly more in place. It may have been interpolated here merely in order to supply a beginning to the chapter.

2. When in difficult country, do not encamp. In country where high roads intersect, join hands with your allies. Do not linger in dangerously isolated positions.

   Chang Yü [defines difficult country as] situated across the frontier, in hostile territory. Li Ch'üan says it is "country in which there are no springs or wells, flocks or herds, vegetables or firewood"; Chia Lin, "one of gorges, chasms and precipices, without a road by which to advance."

In hemmed-in situations, you must resort to stratagem. In desperate position, you must fight.

3. There are roads which must not be followed,

"Especially those leading through narrow defiles," says Li Ch'üan, "where an ambush is to be feared."

armies which must not be attacked,

More correctly, perhaps, "there are times when an army must not be attacked." Ch'ên Hao says: "When you see your way to obtain a rival advantage, but are powerless to inflict a real defeat, refrain from attacking, for fear of overtaxing your men's strength."

towns which must not be besieged,

Cf., ch. III, sect. 4. Ts'ao Kung gives an interesting illustration from his own experience. When invading the territory of Hsu-chou, he ignored the city of Hua-pi, which lay directly in his path, and pressed on into the heart of the country. This excellent strategy was rewarded by the subsequent capture of no fewer than fourteen important district cities. Chang Yü says: "No town should be attacked which, if taken, cannot be held, or if left alone, will not cause any trouble." Hsun Ying, when urged to attack Pi-yang, replied: "The city is small and well-fortified; even if I succeed intaking it, it will be no great feat of arms; whereas if I fail, I shall make myself a laughing-stock." In the seventeenth century, sieges still formed a large proportion of war. It was Turenne who directed attention to the importance of marches, countermarches and maneuvers. He said: "It is a great mistake to waste men in taking a town when the same expenditure of soldiers will gain a province."

positions which must not be contested, commands of the sovereign which must not be obeyed.

This is a hard saying for the Chinese, with their reverence for

兵
法

authority, and Wei Liao Tzŭ (quoted by Tu Mu) is moved to exclaim: "Weapons are baleful instruments, strife is antagonistic to virtue, a military commander is the negation of civil order!" The unpalatable fact remains, however, that even Imperial wishes must be subordinated to military necessity. . . . Ho Shih says: "Even if it be your sovereign's command to encamp in difficult country, linger in isolated positions, etc., you must not do so." The theory is perhaps a little too ingenious to be accepted with confidence.

4. **The general who thoroughly understands the advantages that accompany variation of tactics knows how to handle his troops.**

5. **The general who does not understand these, may be well acquainted with the configuration of the country, yet he will not be able to turn his knowledge to practical account.**

Literally, "get the advantage of the ground," which means not only securing good positions, but availing oneself of natural advantages in every possible way. Chang Yü says: "Every kind of ground is characterized by certain natural features, and also gives scope for a certain variability of plan. How it is possible to turn these natural features to account unless topographical knowledge is supplemented by versatility of mind?"

6. **So, the student of war who is unversed in the art of varying his plans, even though he be acquainted with the Five Advantages, will fail to make the best use of his men.**

Chia Lin . . . tells us that these imply five obvious and generally advantageous lines of action, namely: "if a certain road is short, it must be followed; if an army is isolated, it must be attacked; if a town is in a parlous condition, it must be besieged; if a position can be stormed, it must be attempted; and if consistent with military operations, the ruler's commands must be obeyed." But there are circumstances which sometimes forbid a general to use these advantages.

For instance, "a certain road may be the shortest way for him, but if he knows that it abounds in natural obstacles, or that the enemy has laid an ambush on it, he will not follow that road. A hostile force may be open to attack, but if he knows that it is hard-pressed and likely to fight with desperation, he will refrain from striking," and so on.

7. **Hence in the wise leader's plans, considerations of advantage and of disadvantage will be blended together.**

"Whether in an advantageous position or a disadvantageous one," says Ts'ao Kung, "the opposite state should be always present to your mind."

8. **If our expectation of advantage be tempered in this way, we may succeed in accomplishing the essential part of our schemes.**

[Tu Mu says]: "If we wish to wrest an advantage from the enemy, we must not fix our minds on that alone, but allow for the possibility of the enemy also doing some harm to us, and let this enter as a factor into our calculations."

9. **If, on the other hand, in the midst of difficulties we are always ready to seize an advantage, we may extricate ourselves from misfortune.**

Tu Mu says: "If I wish to extricate myself from a dangerous position, I must consider not only the enemy's ability to injure me, but also my own ability to gain an advantage over the enemy. If in my counsels these two considerations are properly blended, I shall succeed in liberating myself. . . . For instance; if I am surrounded by the enemy and only think of effecting an escape, the nervelessness of my policy will incite my adversary to pursue and crush me; it would be far better to encourage my men to deliver a bold counter-attack, and use the advantage thus gained to free myself from the enemy's toils." See the story of Ts'ao Ts'ao, ch. VII, sect. 35.

孫
子

10. **Reduce the hostile chiefs by inflicting damage on them;**

Chia Lin enumerates several ways of inflicting this injury, some of which would only occur to the Oriental mind:—"Entice away the enemy's best and wisest men, so that he may be left without counselors. Introduce traitors into his country, that the government policy may be rendered futile. Foment intrigue and deceit, and thus sow dissension between the ruler and his ministers. By means of every artful contrivance, cause deterioration amongst his men and waste of his treasure. Corrupt his morals by insidious gifts leading him into excess. Disturb and unsettle his mind by presenting him with lovely women." Chang Yü (after Wang Hsi) [makes a different interpretation of Sun Tzŭ here]: "Get the enemy into a position where he must suffer injury, and he will submit of his own accord."

make trouble for them, and keep them constantly engaged;

[L]iterally, "make servants of them." Tu Yu says "prevent them from having any rest."

hold out specious allurements, and make them rush to any given point.

Mêng Shih's note contains an excellent [interpretation] "cause them to forget *pien* (the reasons for acting otherwise than on their first impulse), and hasten in our direction."

11. **The art of war teaches us to rely not on the likelihood of the enemy's not coming, but on our own readiness to receive him; not on the chance of his not attacking, but rather on the fact that we have made our position unassailable.**

12. **There are five dangerous faults which may affect a general: (1) Recklessness, which leads to destruction;**

"Bravery without forethought," as Ts'ao Kung analyzes it, which causes a man to fight blindly and desperately like a mad bull. Such an opponent, says Chang Yü, "must not be encountered with brute force, but may be lured into an ambush and slain." *Cf.* Wu Tzŭ, ch. IV. *ad init.*: "In estimating the character of a general, men are wont to pay exclusive attention to his courage, forgetting that courage is only one out of many qualities which a general should possess. The merely brave man is prone to fight recklessly; and he who fights recklessly, without any perception of what is expedient, must be condemned." Ssŭ-ma Fa, too, make the incisive remark: "Simply going to one's death does not bring about victory."

(2) cowardice, which leads to capture;

[E]xplained by Ts'ao Kung of the man "whom timidity prevents from advancing to seize an advantage," and Wang Hsi adds "who is quick to flee at the sight of danger." Mêng Shih gives the closer paraphrase "he who is bent on returning alive," this is, the man who will never take a risk. But, as Sun Tzŭ knew, nothing is to be achieved in war unless you are willing to take risks. T'ai Kung said: "He who lets an advantage slip will subsequently bring upon himself real disaster." In AD 404, Liu Yu pursued the rebel Huan Hsuan up the Yangtsze and fought a naval battle with him at the island of Ch'êng-hung. The loyal troops numbered only a few thousands, while their opponents were in great force. But Huan Hsuan, fearing the fate which was in store for him should he be overcome, had a light boat made fast to the side of his war-junk, so that he might escape, if necessary, at a moment's notice. The natural result was that the fighting spirit of his soldiers was utterly quenched, and when the loyalists made an attack from windward with fireships, all striving with the utmost ardor to be first in the fray, Huan Hsuan's forces were routed, had to burn all their baggage and fled for two days and nights without stopping. Chang Yü tells a somewhat similar story of Chao Ying-Ch'i, a general of the Chin State who during a battle with the army of Ch'u in 597 BC had a boat kept in readiness for him on the river, wishing in case of defeat to be the first to get across.

(3) a hasty temper, which can be provoked by insults;

Tu Mu tells us that Yao Hsing, when opposed in AD 357 by Huang Mei, Teng Ch'iang and others shut himself up behind his walls and refused to fight. Teng Ch'iang said: "Our adversary is of a choleric temper and easily provoked; let us make constant sallies and break down his walls, then he will grow angry and come out. Once we can bring his force to battle, it is doomed to be our prey." This plan was acted upon, Yao Hsiang came out to fight, was lured as far as San-Yüan by the enemy's pretended flight, and finally attacked and slain.

(4) a delicacy of honor which is sensitive to shame;

This need not be taken to mean that a sense of honor is really a defect in a general. What Sun Tzŭ condemns is rather an exaggerated sensitiveness to slanderous reports, the thin-skinned man who is stung by opprobrium, however undeserved. Mei Yao-ch'ên truly observes, though somewhat paradoxically: "The seek after glory should be careless of public opinion."

(5) over-solicitude for his men, which exposes him to worry and trouble.

Here again, Sun Tzŭ does not mean that the general is to be careless of the welfare of his troops. All he wishes to emphasize is the danger of sacrificing any important military advantage to the immediate comfort of his men. This is a shortsighted policy, because in the long run the troops will suffer more from the defeat, or, at best, the prolongation of the war, which will be the consequence. A mistaken feeling of pity will often induce a general to relieve a beleaguered city, or to reinforce a hard-pressed detachment, contrary to his military instincts. It is now generally admitted that our repeated efforts to relieve Ladysmith in the South African War were so many strategical blunders which defeated their own purpose. And in the end, relief came through the very man who started out with the distinct resolve no longer to subordinate the interests of the whole to sentiment in favor of a part. An old soldier of one of our generals who failed most conspicuously in this war, tried once, I remember, to defend him to me on the ground that he was always "so good to his men." By this plea, had he but known it, he was only condemning him out of Sun Tzŭ's mouth.

13. **These are the five besetting sins of a general, ruinous to the conduct of war.**

14. **When an army is overthrown and its leader slain, the cause will surely be found among these five dangerous faults. Let them be a subject of meditation.**

兵
法

THE ARMY
ON THE MARCH

The contents of this interesting chapter are better indicated in sect. 1 than in this heading.

1. **Sun Tzŭ said: We come now to the question of encamping the army, and observing signs of the enemy. Pass quickly over mountains, and keep in the neighborhood of valleys.**

   [According to Tu Mu,] the idea is, not to linger among barren uplands, but to keep close to supplies of water and grass. . . . *Cf.* Wu Tzŭ: "Abide not in natural ovens," *i.e.*, "the openings of valleys." Chang Yü tells the following anecdote: "Wu-tu Ch'iang was a robber captain in the time of the Later Han, and Ma Yüan was sent to exterminate his gang. Ch'iang having found a refuge in the hills, Ma Yüan made no attempt to force a battle, but seized all the favorable positions commanding supplies of water and forage. Ch'iang was soon in such a desperate plight for want of provisions that he was forced to make a total surrender. He did not know the advantage of keeping in the neighborhood of valleys."

2. **Camp in high places,**

   Not on high hills, but on knolls or hillocks elevated above the surrounding country.

   facing the sun.

   Tu Mu takes this to mean "facing south," and Ch'ên Hao "facing east." *Cf. infra*, sects. 11, 13.

Do not climb heights in order to fight. So much for mountain warfare.

3. **After crossing a river, you should get far away from it.**

"In order to tempt the enemy to cross after you," according to Ts'ao Kung, and also, says Chang Yü, "in order not to be impeded in your evolutions." The *T'ung Tien* reads, "If *the enemy* crosses a river," etc. But in view of the next sentence, this is almost certainly an interpolation.

4. **When an invading force crosses a river in its onward march, do not advance to meet it in mid-stream. It will be best to let half the army get across, and then deliver your attack.**

Li Ch'üan alludes to the great victory won by Han Hsin over Lung Chü at the Wei River. In the *Ch'ien Han Shu*, the battle is described as follows: "The two armies were drawn up on opposite sides of the river. In the night, Han Hsin ordered his men to take some ten thousand sacks filled with sand and construct a dam higher up. Then, leading half his army across, he attacked Lung Chü; but after a time, pretending to have failed in his attempt, he hastily withdrew to the other bank. Lung Chü was much elated by this unlooked for success, and exclaiming: "I felt sure that Han Hsin was really a coward!" he pursued him and began crossing the river in his turn. Han Hsin now sent a party to cut open the sandbags, thus releasing a great volume of water, which swept down and prevented the greater portion of Lung Chü's army from getting across. He then turned upon the force which had been cut off, and annihilated it, Lung Chü himself being amongst the slain. The rest of the army, on the further bank, also scattered and fled in all directions."

兵
法

5. **If you are anxious to fight, you should not go to meet the invader near a river which he has to cross.**

For fear of preventing his crossing.

6. **Moor your craft higher up than the enemy, and facing the sun.**

See *supra*, sect. 2. The repetition of these words in connection with water is very awkward. Chang Yü has the note: "Said either of troops marshaled on the river-bank, or of boats anchored in the stream itself; in either case it is essential to be higher than the enemy and facing the sun." The other commentators are not at all explicit.

Do not move up-stream to meet the enemy.

Tu Mu says: "As water flows downwards, we must not pitch our camp on the lower reaches of a river, for fear the enemy should open the sluices and sweep us away in a flood.... Chu-ko Wu-hou has remarked that 'in river warfare we must not advance against the stream,' which is as much as to say that our fleet must not be anchored below that of the enemy, for then they would be able to take advantage of the current and make short work of us." There is also the danger, noted by other commentators, that the enemy may throw poison on the water to be carried down to us.

So much for river warfare.

7. **In crossing salt-marshes, your sole concern should be to get over them quickly, without any delay.**

Because of the lack of fresh water, the poor quality of the herbage, and last but not least, because they are low, flat, and exposed to attack.

8. **If forced to fight in a salt-marsh, you should have water and grass near you, and get your back to a clump of trees.**

Li Ch'üan remarks that the ground is less likely to be treacherous where there are trees, while Tu Mu says that they will serve to protect the rear.

So much for operations in salt-marshes.

9. **In dry, level country, take up an easily accessible position with rising ground to your right and on your rear,**

Tu Mu quotes T'ai Kung as saying: "An army should have a stream or a marsh on its left, and a hill or tumulus on its right."

so that the danger may be in front, and safety lie behind. So much for campaigning in flat country.

10. **These are the four useful branches of military knowledge**

Those, namely, concerned with (1) mountains, (2) rivers, (3) marshes, and (4) plains.

which enabled the Yellow Emperor to vanquish four several sovereigns.

兵
法

11. **All armies prefer high ground to low,**

"High Ground," says Mei Yao-ch'ên, "is not only more agreement and salubrious, but more convenient from a military point of view; low ground is not only damp and unhealthy, but also disadvantageous for fighting."

and sunny places to dark.

12. **If you are careful of your men,**

Ts'ao Kung says: "Make for fresh water and pasture, where you can turn out your animals to graze."

and camp on hard ground, the army will be free from disease of every kind,

Chang Yü says: "The dryness of the climate will prevent the outbreak of illness."

and this will spell victory.

13. **When you come to a hill or a bank, occupy the sunny side, with the slope on your right rear. Thus you will at once act for the benefit of your soldiers and utilize the natural advantages of the ground.**

14. **When, in consequence of heavy rains up-country, a river which you wish to ford is swollen and flecked with foam, you must wait until it subsides.**

15. **Country in which there are precipitous cliffs with torrents running between, deep natural hollows,**

    [E]xplained as "places enclosed on every side by steep banks, with pools of water at the bottom."

    confined places,

    "[N]atural pens or prisons" explained as "places surrounded by precipices on three sides—easy to get into, but hard to get out of."

    tangled thickets,

    [E]xplained as "places covered with such dense undergrowth that spears cannot be used."

    quagmires

    Explained as "low-lying places, so heavy with mud as to be impassable for chariots and horsemen."

and crevasses,

[E]xplained by Mei Yao-ch'ên as "a narrow difficult way between bee-tling cliffs." . . . Tu Mu's note is "ground covered with trees and rocks, and intersected by numerous ravines and pitfalls." This is very vague, but Chia Lin explains it clearly enough as a defile or narrow pass . . . and Chang Yü takes much the same view. On the whole, the weight of the commentators certainly inclines to the rendering "defile."

should be left with all possible speed and not approached.

16. **While we keep away from such places, we should get the enemy to approach them; while we face them, we should let the enemy have them on his rear.**

17. **If in the neighborhood of your camp there should be any hilly country, ponds surrounded by aquatic grass, hollow basins filled with reeds, or woods with thick undergrowth, they must be carefully routed out and searched; for these are places where men in ambush or insidious spies are likely to be lurking.**

    Chang Yü [notes]: "We must also be on our guard against traitors who may lie in close covert, secretly spying out our weaknesses and overhearing our instructions."

18. **When the enemy is close at hand and remains quiet, he is relying on the natural strength of his position.**

    Here begin Sun Tzǔ's remarks on the reading of signs, much of which is so good that it could almost be included in a modern manual like Gen. Baden-Powell's *Aids to Scouting.*

19. **When he keeps aloof and tries to provoke a battle, he is anxious for the other side to advance.**

Probably because we are in a strong position from which he wishes to dislodge us. "If he came close up to us," says Tu Mu, "and tried to force a battle, he would seem to despise us, and there would be less probability of our responding to the challenge."

20. **If his place of encampment is easy of access, he is tendering a bait.**

21. **Movement amongst the trees of a forest shows that the enemy is advancing.**

Ts'ao Kung explains this as "felling trees to clear a passage," and Chang Yü says: "Every man sends out scouts to climb high places and observe the enemy. If a scout sees that the trees of a forest are moving and shaking, he may know that they are being cut down to clear a passage for the enemy's march."

The appearance of a number of screens in the midst of thick grass means that the enemy wants to make us suspicious.

Tu Yu's explanation, borrowed from Ts'ao Kung's, is as follows: "The presence of a number of screens or sheds in the midst of thick vegetation is a sure sign that the enemy has fled and, fearing pursuit, has constructed these hiding-places in order to make us suspect an ambush." It appears that these "screens" were hastily knotted together out of any long grass which the retreating enemy happened to come across.

22. **The rising of birds in their flight is the sign of an ambuscade.**

Chang Yü's explanation is doubtless right: "When birds that are flying along in a straight line suddenly shoot upwards, it means that soldiers are in ambush at the spot beneath."

Startled beasts indicate that a sudden attack is coming.

23. **When there is dust rising in a high column, it is the sign of chariots advancing; when the dust is low, but spread over a wide area, it betokens the approach of infantry.**

[The text that translates as] "high and sharp," or rising to a peak, is of course somewhat exaggerated as applied to dust. The commentators explain the phenomenon by saying that horses and chariots, being heavier than men, raise more dust, and also follow one another in the same wheel-track, whereas foot-soldiers would be marching in ranks, many abreast. According to Chang Yü, "every army on the march must have scouts some way in advance, who on sighting dust raised by the enemy, will gallop back and report it to the commander-in-chief." *Cf.* Gen. Baden-Powell [in *Aids to Scouting*]: "As you move along, say, in a hostile country, your eyes should be looking afar for the enemy or any signs of him: figures, dust rising, birds getting up, glitter of arms, etc."

When it branches out in different directions, it shows that parties have been sent to collect firewood. A few clouds of dust moving to and fro signify that the army is encamping.

Chang Yü says: "In apportioning the defenses for a cantonment, light horse will be sent out to survey the position and ascertain the weak and strong points all along its circumference. Hence the small quantity of dust and its motion."

24. **Humble words and increased preparations are signs that the enemy is about to advance.**

"As though they stood in great fear of us," says Tu Mu. "Their object is to make us contemptuous and careless, after which they will attack us." Chang Yü alludes to the story of T'ien Tan of the Ch'i-mo against the Yen forces, led by Ch'i Chieh. In the *Shih Chi* we read: "T'ien Tan openly said: 'My only fear is that the Yen army may cut off the noses of their Ch'i prisoners and place them in the front rank to fight against us; that would be the undoing of our city.' The other side being informed of this speech, at once acted on the suggestion; but those within the city were enraged at seeing their fellow-countrymen thus mutilated, and fearing only lest they should fall into the enemy's hands, were nerved to defend themselves more obstinately than ever. Once again T'ien Tan sent back converted spies who reported these words to the enemy: 'What I dread most is that the men of Yen may dig up the ancestral tombs outside the town, and by inflicting this indignity on our forefathers cause us to become faint-hearted.' Forthwith the besiegers dug up all the graves and burned the corpses lying in them. And the inhabitants of Chi-mo, witnessing the outrage from the city-walls, wept passionately and were all impatient to go out and fight, their fury being increased tenfold. T'ien Tan knew then that his soldiers were ready for any enterprise. But instead of a sword, he himself took a mattock in his hands, and ordered others to be distributed amongst his best warriors, while the ranks were filled up with their wives and concubines. He then served out all the remaining rations and bade his men eat their fill. The regular soldiers were told to keep out of sight, and the walls were manned with the old and weaker men and with women. This done, envoys were dispatched to the enemy's camp to arrange terms of surrender, whereupon the Yen army began shouting for joy. T'ien Tan also collected 20,000 ounces of silver from the people, and got the wealthy citizens of Chi-mo to send it to the Yen general with the prayer that, when the town capitulated, he would not allow their homes to be plundered or their women to be maltreated. Ch'i Chieh, in high good humor, granted their prayer; but his army now became increasingly slack and careless. Meanwhile, T'ien Tan got together a thousand oxen, decked them with pieces of red silk, painted their bodies, dragon-like, with colored stripes, and fastened sharp blades on their horns and well-greased rushes on their tails. When night came on, he lighted the ends of the rushes, and drove the oxen through a number of holes which he had pierced in the walls, backing them up with a force of 5,000 picked warriors. The animals,

兵
法

maddened with pain, dashed furiously into the enemy's camp where they caused the utmost confusion and dismay; for their tails acted as torches, showing up the hideous pattern on their bodies, and the weapons on their horns killed or wounded any with whom they came into contact. In the meantime, the band of 5,000 had crept up with gags in their mouths, and now threw themselves on the enemy. At the same moment a frightful din arose in the city itself, all those that remained behind making as much noise as possible by banging drums and hammering on bronze vessels, until heaven and earth were convulsed by the uproar. Terror-stricken, the Yen army fled in disorder, hotly pursued by the men of Ch'i, who succeeded in slaying their general Ch'i Chien. . . . The result of the battle was the ultimate recovery of some seventy cities which had belonged to the Ch'i State."

孫
子

Violent language and driving forward as if to the attack are signs that he will retreat.

25. **When the light chariots come out first and take up a position on the wings, it is a sign that the enemy is forming for battle.**

26. **Peace proposals unaccompanied by a sworn covenant indicate a plot.**

27. **When there is much running about**

Every man hastening to his proper place under his own regimental banner.

and the soldiers fall into rank, it means that the critical moment has come.

28. **When some are seen advancing and some retreating, it is a lure.**

29. When the soldiers stand leaning on their spears, they are faint from want of food.

30. If those who are sent to draw water begin by drinking themselves, the army is suffering from thirst.

     As Tu Mu remarks: "One may know the condition of a whole army from the behavior of a single man."

31. If the enemy sees an advantage to be gained and makes no effort to secure it, the soldiers are exhausted.

32. If birds gather on any spot, it is unoccupied.

     A useful fact to bear in mind when, for instance, as Ch'ên Hao says, the enemy has secretly abandoned his camp.

     Clamor by night betokens nervousness.

33. If there is disturbance in the camp, the general's authority is weak. If the banners and flags are shifted about, sedition is afoot. If the officers are angry, it means that the men are weary.

     Tu Mu understands the sentence to read . . . "If all the officers of an army are angry with their general, it means that they are broken with fatigue" (owing to the exertions which he has demanded from them).

34. When an army feeds its horses with grain and kills its cattle for food,

     In the ordinary course of things, the men would be fed on grain and the horses chiefly on grass.

兵
法

and when the men do not hang their cooking-pots over the camp-fires, showing that they will not return to their tents, you may know that they are determined to fight to the death.

I may quote here the illustrative passage from the *Hou Han Shu*, given in abbreviated form by the *P'ei Wên Yün Fu*: "The rebel Wang Kuo of Liang was besieging the town of Ch'ên-ts'ang, and Huang-fu Sung, who was in supreme command, and Tung Cho were sent out against him. The latter pressed for hasty measures, but Sung turned a deaf ear to his counsel. At last the rebels were utterly worn out, and began to throw down their weapons of their own accord. Sung was not advancing to the attack, but Cho said: 'It is a principle of war not to pursue desperate men and not to press a retreating host.' Sung answered: 'That does not apply here. What I am about to attack is a jaded army, not a retreating host; with disciplined troops I am falling on a disorganized multitude, not a band of desperate men.' Thereupon he advances to the attack unsupported by his colleague, and routed the enemy, Wang Kuo being slain."

孫
子

35. **The sight of men whispering together in small knots or speaking in subdued tones points to disaffection amongst the rank and file.**

36. **Too frequent rewards signify that the enemy is at the end of his resources;**

Because, when an army is hard pressed, as Tu Mu says, there is always a fear of mutiny, and lavish rewards are given to keep the men in good temper.

too many punishments betray a condition of dire distress.

Because in such case discipline becomes relaxed, and unwonted severity is necessary to keep the men to their duty.

37. **To begin by bluster, but afterwards to take fright at the enemy's numbers, shows a supreme lack of intelligence.**

I follow the interpretation of Ts'ao Kung...also adopted by Li Ch'üan, Tu Mu, and Chang Yü. Another possible meaning set forth by Tu Yu, Chia Lin, Mei Tao-Ch'ên and Wang Hsi, is: "The general who is first tyrannical towards his men, and then in terror lest they should mutiny, etc." This would connect the sentence with what went before about rewards and punishments.

38. **When envoys are sent with compliments in their mouths, it is a sign that the enemy wishes for a truce.**

Tu Mu says: "If the enemy open friendly relations by sending hostages, it is a sign that they are anxious for an armistice, either because their strength is exhausted or for some other reason." But it hardly needs a Sun Tzŭ to draw such an obvious inference.

39. **If the enemy's troops march up angrily and remain facing ours for a long time without either joining battle or taking themselves off again, the situation is one that demands great vigilance and circumspection.**

Ts'ao Kung says a maneuver of this sort may be only a ruse to gain time for an unexpected flank attack or the laying of an ambush.

40. **If our troops are no more in number than the enemy, that is amply sufficient; it only means that no direct attack can be made.**

Literally, "no martial advance." That is to say, *chêng* tactics and frontal attacks must be eschewed, and stratagem resorted to instead.

What we can do is simply to concentrate all our available strength, keep a close watch on the enemy, and obtain reinforcements.

This is an obscure sentence, and none of the commentators succeed in squeezing very good sense out of it. . . . I follow Li Ch'üan, who appears to offer the simplest explanation: "Only the side that gets more men will win." . . . Fortunately we have Chang Yü to expound its meaning to us in language which is lucidity itself: "When the numbers are even, and no favorable opening presents itself, although we may not be strong enough to deliver a sustained attack, we can find additional recruits amongst our sutlers and camp-followers, and then, concentrating our forces and keeping a close watch on the enemy, contrive to snatch the victory. But we must avoid borrowing foreign soldiers to help us." He then quotes from Wei Liao Tzŭ: "The nominal strength of mercenary troops may be 100,000, but their real value will be not more than half that figure." According to this interpretation, [words in the text mean] "to get recruits," not from outside, but from the rag-tag and bobtail which follow in the wake of a large army. This does not sound a very soldierly suggestion, and I feel convinced that it is not what Sun Tzŭ meant. Chia Lin, on the other hand, takes the words in a different sense altogether, namely "to conquer the enemy." (*Cf.* ch. I, sect. 20.)

41. **He who exercises no forethought but makes light of his opponents is sure to be captured by them.**

     [Ch'ên Hao, quoting from the *Tso Chuan*, says]: "If bees and scorpions carry poison, how much more will a hostile state! Even a puny opponent, then, should not be treated with contempt."

42. **If soldiers are punished before they have grown attached to you, they will not prove submissive; and, unless submissive, then will be practically useless. If, when the soldiers have become attached to you, punishments are not enforced, they will still be useless.**

43. **Therefore soldiers must be treated in the first instance with humanity, but kept under control by means of iron discipline.**

    Yen Tzŭ (493 BC) said of Ssŭ-ma Jang-chu: "His civil virtues endeared him to the people; his martial prowess kept his enemies in awe." *Cf.* Wu Tzŭ: "The ideal commander unites culture with a warlike temper; the profession of arms requires a combination of hardness and tenderness."

    This is a certain road to victory.

44. **If in training soldiers commands are habitually enforced, the army will be well-disciplined; if not, its discipline will be bad.**

45. **If a general shows confidence in his men but always insists on his orders being obeyed,**

    [Tu Mu] says: "A general ought in time of peace to show kindly confidence in his men and also make his authority respected, so that when they come to face the enemy, orders may be executed and discipline maintained, because they all trust and look up to him." What Sun Tzŭ has said in sect. 44, however, would lead one rather to expect something like this: "If a general is always confident that his orders will be carried out," etc.

    兵
    法

    the gain will be mutual.

    Chang Yü says: "The general has confidence in the men under his command, and the men are docile, having confidence in him. Thus the gain is mutual." He quotes a pregnant sentence from Wei Liao Tzŭ: "The art of giving orders is not to try to rectify minor blunders and not to be swayed by petty doubts." Vacillation and fussiness are the surest means of sapping the confidence of an army.

TERRAIN

Only about a third of the chapter, comprising sects. 1–13, deals with [terrain], the subject being more fully treated in ch. XI. The "six calamities" are discussed in sects. 14–20, and the rest of the chapter is again a mere string of desultory remarks, though not less interesting, perhaps, on that account.

1. Sun Tzŭ said: We may distinguish six kinds of terrain, to wit:

(1) Accessible ground;

Mei Yao-ch'ên says: "plentifully provided with roads and means of communications."

(2) entangling ground;

The same commentator says: "Net-like country, venturing into which you become entangled."

(3) temporizing ground;

Ground which allows you to "stave off" or "delay."

(4) narrow passes; (5) precipitous heights; (6) positions at a great distance from the enemy.

It is hardly necessary to point out the faultiness of this classification. A strange lack of logical perception is shown in the Chinaman's unquestioning acceptance of glaring cross-divisions such as the above.

2. Ground which can be freely traversed by both sides is called accessible.

3. With regard to ground of this nature, be before the enemy in occupying the raised and sunny spots, and carefully guard your line of supplies.

> The general meaning of the last phrase is doubtlessly, as Tu Yu says, "not to allow the enemy to cut your communications." In view of Napoleon's dictum, "the secret of war lies in the communications" [*Pensées de Napoléon 1er*, no. 47], we could wish that Sun Tzŭ had done more than skirt the edge of this important subject here and in ch. I, sect. 10 and ch. VII, sect. 11. Col. Henderson says [in ch. 2 of *The Science of War*]: "The line of supply may be said to be as vital to the existence of an army as the heart to the life of a human being. Just as the duelist who finds his adversary's point menacing him with certain death, and his own guard astray, is compelled to conform to his adversary's movements, and to content himself with warding off his thrusts, so the commander whose communications are suddenly threatened finds himself in a false position, and he will be fortunate if he has not to change all his plans, to split up his force into more or less isolated detachments, and to fight with inferior numbers on ground which he has not had time to prepare, and where defeat will not be an ordinary failure, but will entail the ruin or surrender of his whole army."

Then you will be able to fight with advantage.

4. Ground which can be abandoned but is hard to re-occupy is called *entangling*.

5. From a position of this sort, if the enemy is unprepared, you may sally forth and defeat him. But if the enemy is prepared for your coming, and you fail to defeat him, then, return being impossible, disaster will ensue.

6. **When the position is such that neither side will gain by making the first move, it is called *temporizing* ground.**

[Tu Yu says]: "Each side finds it inconvenient to move, and the situation remains at a deadlock."

7. **In a position of this sort, even though the enemy should offer us an attractive bait,**

Tu Yu says, "turning their backs on us and pretending to flee." But this is only one of the lures which might induce us to quit our position.

it will be advisable not to stir forth, but rather to retreat, thus enticing the enemy in his turn; then, when part of his army has come out, we may deliver our attack with advantage.

8. **With regard to *narrow passes*, if you can occupy them first, let them be strongly garrisoned and await the advent of the enemy.**

Because then, as Tu Yu observes, "the initiative will lie with us, and by making sudden and unexpected attacks we shall have the enemy at our mercy."

9. **Should the army forestall you in occupying a pass, do not go after him if the pass is fully garrisoned, but only if it is weakly garrisoned.**

10. **With regard to *precipitous heights*, if you are beforehand with your adversary, you should occupy the raised and sunny spots, and there wait for him to come up.**

Ts'ao Kung says: "The particular advantage of securing heights and defiles is that your actions cannot then be dictated by the enemy." (For the enunciation of the grand principle alluded to, see *ch. VI*, sect. 2.) Chang Yü tells the following anecdote of P'ei Hsing-chien (AD 619–682), who was sent on a punitive expedition against the Turkic tribes. "At night he pitched his camp as usual, and it had already been completely fortified by wall and ditch, when suddenly he gave orders that the army should shift its quarters to a hill near by. This was highly displeasing to his officers, who protested loudly against the extra fatigue which it would entail on the men. P'ei Hsing-chien, however, paid no heed to their remonstrances and had the camp moved as quickly as possible. The same night, a terrific storm came on, which flooded their former place of encampment to the depth of over twelve feet. The recalcitrant officers were amazed at the sight, and owned that they had been in the wrong. 'How did you know what was going to happen?' they asked. P'ei Hsing-chien replied: 'From this time forward be content to obey orders without asking unnecessary questions.' From this it may be seen," Chang Yü continues, "that high and sunny places are advantageous not only for fighting, but also because they are immune from disastrous floods."

11. **If the enemy has occupied them before you, do not follow him, but retreat and try to entice him away.**

The turning point of Li Shih-min's campaign in AD 621, against the two rebels, Tou Chien-tê, King of Hsia, and Wang Shih-ch'ung, Prince of Chêng, was his seizure of the heights of Wu-lao, in spite of which Tou Chien-tê persisted in his attempt to relieve his ally in Lo-yang, was defeated and taken prisoner.

12. **If you are situated at a great distance from the enemy, and the strength of the two armies is equal, it is not easy to provoke a battle,**

The point of course, is that we must not think of undertaking a long and wearisome march, at the end of which "we should be exhausted and our adversary fresh and keen."

and fighting will be to your disadvantage.

13. **These six are the principles connected with Earth.**

    Or perhaps, "the principles relating to ground." See, however, ch. I, sect. 8.

    The general who has attained a responsible post must be careful to study them.

14. **Now an army is exposed to six several calamities, not arising from natural causes, but from faults for which the general is responsible. These are: (1) Flight; (2) insubordination; (3) collapse; (4) ruin; (5) disorganization; (6) rout.**

15. **Other conditions being equal, if one force is hurled against another ten times its size, the result will be the *flight* of the former.**

16. **When the common soldiers are too strong and their officers too weak, the result is *insubordination*.**

    Tu Mu cites the unhappy case of T'ien Pu . . . , who was sent to Wei in AD 821 with orders to lead an army against Wang T'ing-ts'ou. But the whole time he was in command, his soldiers treated him with the utmost contempt, and openly flouted his authority by riding about the camp on donkeys, several thousands at a time. T'ien Pu was powerless to put a stop to this conduct, and when, after some months had passed, he made an attempt to engage the enemy, his troops turned tail and dispersed in every direction. After that, the unfortunate man committed suicide by cutting his throat.

    When the officers are too strong and the common soldiers too weak, the result is *collapse*.

    Ts'ao Kung says: "The officers are energetic and want to press on, the common soldiers are feeble and suddenly collapse."

186

17. **When the higher officers are angry and insubordinate, and on meeting the enemy give battle on their own account from a feeling of resentment, before the commander-in-chief can tell whether or no he is in a position to fight, the result is *ruin*.**

    Wang Hsi's note is: "This means, the general is angry without cause, and at the same time does not appreciate the ability of his subordinate officers; thus he arouses fierce resentment and brings an avalanche of ruin upon his head." . . . [B]ut I think Ch'ên Hao is right in his paraphrase "they don't care if it be possible or no."

18. **When the general is weak and without authority; when his orders are not clear and distinct;**

    Wei Liao Tzŭ says: "If the commander gives his orders with decision, the soldiers will not wait to hear them twice; if his moves are made without vacillation, the soldiers will not be in two minds about doing their duty." General Baden-Powell says [in *Aids to Scouting*], italicizing the words: "The secret of getting successful work out of your trained men lies in one nutshell—in the clearness of the instructions they receive.". . . *Cf.* also Wu Tzŭ: "the most fatal defect in a military leader is difference; the worst calamities that befall an army arise from hesitation."

    when there are no fixed duties assigned to officers and men,

    [Tu Mu says]: "Neither officers nor men have any regular routine."

    and the ranks are formed in a slovenly haphazard manner, the result is utter *disorganization*.

19. **When a general, unable to estimate the enemy's strength, allows an inferior force to engage a larger one, or hurls a weak detachment against a powerful one, and neglects to place picked soldiers in the front rank, the result must be *rout*.**

兵
法

Chang Yü paraphrases the latter part of the sentence and continues: "Whenever there is fighting to be done, the keenest spirits should be appointed to serve in the front ranks, both in order to strengthen the resolution of our own men and to demoralize the enemy."

20. **These are six ways of courting defeat, which must be carefully noted by the general who has attained a responsible post.**

See *supra*, sect. 13.

21. **The natural formation of the country is the soldier's best ally;**

Ch'ên Hao says: "The advantages of weather and season are not equal to those connected with ground."

but a power of estimating the adversary, of controlling the forces of victory, and of shrewdly calculating difficulties, dangers and distances, constitutes the test of a great general.

22. **He who knows these things, and in fighting puts his knowledge into practice, will win his battles. He who knows them not, nor practices them, will surely be defeated.**

23. **If fighting is sure to result in victory, then you must fight, even though the ruler forbid it; if fighting will not result in victory, then you must not fight even at the ruler's bidding.**

Huang Shih-kung of the Ch'in dynasty, who is said to have been the patron of Chang Liang and to have written the [*San Lueh*], has these words attributed to him: "The responsibility of setting an army in motion must devolve on the general alone; if advance and retreat are controlled from the Palace, brilliant results will hardly be achieved. Hence the god-like ruler and the enlightened monarch are content to play a humble part in furthering their country's cause (*literally*, kneel down to push the chariot wheel)." This means that "in matters lying outside the *zenana*, the decision of the military commander must be absolute." Chang Yü also quotes the saying: "Decrees of the Son of Heaven do not penetrate the walls of a camp."

24. **The general who advances without coveting fame and retreats without fearing disgrace,**

It was Wellington, I think, who said that the hardest thing of all for a soldier is to retreat.

whose only thought is to protect his country and do good service for his sovereign, is the jewel of the kingdom.

A noble presentiment, in few words, of the Chinese "happy warrior." Such a man, says Ho Shih, "even if he had to suffer punishment, would not regret his conduct."

25. **Regard your soldiers as your children, and they will follow you into the deepest valleys; look upon them as your own beloved sons, and they will stand by you even unto death.**

Cf. ch. I, sect. 6. In this connection, Tu Mu draws for us an engaging picture of the famous general Wu Ch'i, from whose treatise on war I have frequently had occasion to quote: "He wore the same clothes and ate the same food as the meanest of his soldiers, refused to have either a horse to ride or a mat to sleep on, carried his own surplus rations wrapped in a parcel, and shared every hardship with his men. One of his soldiers was suffering from an abscess, and Wu Ch'i himself sucked out the virus. The soldier's mother, hearing this, began wailing and lamenting. Somebody asked her, saying: 'Why do you cry? Your son is only a common soldier, and yet the commander-in-chief himself has sucked the poison from his sore.' The woman replied, 'Many years ago, Lord Wu performed a similar service for my husband, who never left him afterwards, and finally met his death at the hands of the enemy. And now that he has done the same for my son, he too will fall fighting I know not where.'" Li Ch'üan mentions the Viscount of Ch'u, who invaded the small state of Hsiao during the winter. The Duke of Shên said to him: "Many of the soldiers are suffering severely from the cold." So he made a round of the whole army, comforting and encouraging the men; and straightway they felt as if they were clothed in garments lined with floss silk.

26. **If, however, you are indulgent, but unable to make your authority felt; kind-hearted, but unable to enforce your commands; and incapable, moreover, of quelling disorder: then your soldiers must be likened to spoilt children; they are useless for any practical purpose.**

Li Ching once said that if you could make your soldiers afraid of you, they would not be afraid of the enemy. Tu Mu recalls an instance of stern military discipline which occurred in AD 219, when Lü Mêng was occupying the town of Chiang-ling. He had given stringent orders to his army not to molest the inhabitants or take anything from them by force. Nevertheless, a certain officer serving under his banner, who happened to be a fellow-townsman, ventured to appropriate a bamboo hat belonging to one of the people, in order to wear it over his regulation helmet as a protection against the rain. Lu Mêng considered that the fact of his being also a native of Ju-nan should not be allowed to palliate a clear breach of discipline, and accordingly he ordered his summary execution, the tears rolling down his face, however, as he did so. This act of severity filled the army with wholesome awe, and from that time forth even articles dropped in the highway were not picked up.

27. **If we know that our own men are in a condition to attack, but are unaware that the enemy is not open to attack, we have gone only halfway towards victory.**

That is, Ts'ao Kung says, "the issue in this case is uncertain."

28. If we know that the enemy is open to attack, but are unaware that our own men are not in a condition to attack, we have gone only halfway towards victory.

    Cf. ch. III, sect. 13(1).

29. If we know that the enemy is open to attack, and also know that our men are in a condition to attack, but are unaware that the nature of the ground makes fighting impracticable, we have still gone only halfway towards victory.

30. Hence the experienced soldier, once in motion, is never bewildered; once he has broken camp, he is never at a loss.

    The reason being, according to Tu Mu, that he has taken his measures so thoroughly as to ensure victory beforehand. "He does not move recklessly," says Chang Yü, "so that when he does move, he makes no mistakes."

兵
法

31. Hence the saying: If you know the enemy and know yourself, your victory will not stand in doubt; if you know Heaven and know Earth, you may make your victory complete.

    Li Ch'üan sums up as follows: "Given a knowledge of three things— the affairs of men, the seasons of heaven and the natural advantages of earth—victory will invariably crown your battles."

THE NINE SITUATIONS

[Wang Hsi says]: "There are nine military situations, good and bad."

1. Sun Tzŭ said: The art of war recognizes nine varieties of ground: (1) Dispersive ground; (2) facile ground; (3) contentious ground; (4) open ground; (5) ground of intersecting highways; (6) serious ground; (7) difficult ground; (8) hemmed-in ground; (9) desperate ground.

孫
子

2. When a chieftain is fighting in his own territory, it is dispersive ground.

   So called because the soldiers, being near to their homes and anxious to see their wives and children, are likely to seize the opportunity afforded by a battle and scatter in every direction. "In their advance," observes Tu Mu, "they will lack the valor of desperation, and when they retreat, they will find harbors of refuge."

3. When he has penetrated into hostile territory, but to no great distance, it is facile ground.

   Li Ch'üan and Ho Shih say "because of the facility for retreating," and the other commentators give similar explanations. Tu Mu remarks: "When your army has crossed the border, you should burn your boats and bridges, in order to make it clear to everybody that you have no hankering after home."

4. **Ground the possession of which imports great advantage to either side, is contentious ground.**

Ts'ao Kung says: "ground on which the few and the weak can defeat the many and the strong," such as "the neck of a pass," instanced by Li Ch'üan. Thus, Thermopylae was of this classification because the possession of it, even for a few days only, meant holding the entire invading army in check and thus gaining invaluable time. [As Wu Tzǔ writes]: "For those who have to fight in the ratio of one to ten, there is nothing better than a narrow pass." When Lü Kuang was returning from his triumphant expedition to Turkestan in AD 385, and had got as far as I-ho, laden with spoils, Liang Hsi, administrator of Liang-chou, taking advantage of the death of Fu Chien, King of Ch'in, plotted against him and was for barring his way into the province. Yang Han, governor of Kao-ch'ang, counseled him, saying: "Lü Kuang is fresh from his victories in the west, and his soldiers are vigorous and mettlesome. If we oppose him in the shifting sands of the desert, we shall be no match for him, and we must therefore try a different plan. Let us hasten to occupy the defile at the mouth of the Kao-wu pass, thus cutting him off from supplies of water, and when his troops are prostrated with thirst, we can dictate our own terms without moving. Or if you think that the pass I mention is too far off, we could make a stand against him at the I-wu pass, which is nearer. The cunning and resource of Tzǔ-fang himself would be expended in vain against the enormous strength of these two positions." Liang Hsi, refusing to act on this advice, was overwhelmed and swept away by the invader.

5. **Ground on which each side has liberty of movement is open ground.**

[There are various interpretations of the Chinese adjective for this type of ground. Ts'ao Kung says it means] "ground covered with a network of roads," like a chessboard. Another interpretation, suggested by Ho Shih, is "ground on which intercommunication is easy."

兵
法

6. **Ground which forms the key to three contiguous states,**

"Our country adjoining the enemy's and a third country conterminous with both" (Ts'ao Kung). Mêng Shih instances the small principality of Chêng, which was bounded on the north-east by Ch'i, on the west by Chin, and on the south by Ch'u.

so that he who occupies it first has most of the Empire at his command,

The belligerent who holds this dominating position can constrain most of them to become his allies.

is a ground of intersecting highways.

7. **When an army has penetrated into the heart of a hostile country, leaving a number of fortified cities in its rear, it is serious ground.**

Wang Hsi explains the name by saying that "when an army has reached such a point, its situation is serious." Li Ch'üan instances (1) the victorious march of Yo I into the capital of Ch'i in 284 BC, and (2) the attack on Chi'u six years later by the Ch'in general Po Ch'i.

8. **Mountain forests,**

Or simply "forests."

rugged steeps, marshes and fens—all country that is hard to traverse: this is difficult ground.

9. **Ground which is reached through narrow gorges, and from which we can only retire by tortuous paths, so that a small number of the enemy would suffice to crush a large body of our men: this is hemmed-in ground.**

10. **Ground on which we can only be saved from destruction by fighting without delay, is desperate ground.**

The situation, as pictured by Ts'ao Kung, is very similar to the [hemmed-in ground] except that here escape is no longer possible: "A lofty mountain in front, a large river behind, advance impossible, retreat blocked." Ch'ên Hao says: "to be on 'desperate ground' is like sitting in a leaking boat or crouching in a burning house." Tu Mu quotes from Li Ching a vivid description of the plight of an army thus entrapped: "Suppose an army invading hostile territory without the aid of local guides:—it falls into a fatal snare and is at the enemy's mercy. A ravine on the left, a mountain on the right, a pathway so perilous that the horses have to be roped together and the chariots carried in slings, no passage open in front, retreat cut off behind, no choice but to proceed in single file. Then, before there is time to range our soldiers in order of battle, the enemy in overwhelming strength suddenly appears on the scene. Advancing, we can nowhere take a breathing-space; retreating, we have no haven of refuge. We seek a pitched battle, but in vain; yet standing on the defensive, none of us has a moment's respite. If we simply maintain our ground, whole days and months will crawl by; the moment we make a move, we have to sustain the enemy's attacks on front and rear. The country is wild, destitute of water and plants; the army is lacking in the necessaries of life, the horses are jaded and the men worn-out, all the resources of strength and skill unavailing, the pass so narrow that a single man defending it can check the onset of ten thousand; all means of offense in the hands of the enemy, all points of vantage already forfeited by ourselves:—in this terrible plight, even though we had the most valiant soldiers and the keenest of weapons, how could they be employed with the slightest effect?" Students of Greek history may be reminded of the awful close to the Sicilian expedition, and the agony of the Athenians under Nicias and Demonsthenes.

兵
法

11. **On dispersive ground, therefore, fight not. On facile ground, halt not. On contentious ground, attack not.**

But rather let all your energies be bent on occupying the advantageous position first. So Ts'ao Kung. Li Ch'üan and others, however, suppose the meaning to be that the enemy has already forestalled us, so that it would be sheer madness to attack. In the [*Sun Tzŭ Hsu Liu*], when the King of Wu inquires what should be done in this case, Sun Tzŭ replies: "The rule with regard to contentious ground is that those in possession have the advantage over the other side. If a position of this kind is secured first by the enemy, beware of attacking him. Lure him away by pretending to flee—show your banners and sound your drums—make a dash for other places that he cannot afford to lose—trail brushwood and raise a dust—confound his ears and eyes—detach a body of your best troops, and place it secretly in ambuscade. Then your opponent will sally forth to the rescue."

12. **On open ground, do not try to block the enemy's way.**

Because the attempt would be futile, and would expose the blocking force itself to serious risks.

On the ground of intersecting highways, join hands with your allies.

Or perhaps, "form alliances with neighboring states."

13. **On serious ground, gather in plunder.**

On this, Li Ch'üan has the following delicious note: "When an army penetrates far into the enemy's country, care must be taken not to alienate the people by unjust treatment. Follow the example of the Han Emperor Kao Tsu, whose march into Ch'in territory was marked by no violation of women or looting of valuables. (*Nota bene*: this was in 207 BC, and may well cause us to blush for the Christian armies that entered Peking in AD 1900.) Thus he won the hearts of

all. In the present passage, then, I think that the true reading must be, not 'plunder,' but 'do not plunder.'" Alas, I fear that in this instance the worthy commentator's feelings outran his judgment. Tu Mu, at least, has no such illusions. He says: "When encamped on 'serious ground,' there being no inducement as yet to advance further, and no possibility of retreat, one ought to take measures for a protracted resistance by bringing in provisions from all sides, and keep a close watch on the enemy." *Cf.* also ch. II, sect. 9.

In difficult ground, keep steadily on the march.

Or, in the words of ch. VIII, sect. 2, "do not encamp."

14. On hemmed-in ground, resort to stratagem.

Ts'ao Kung says: "Try the effect of some unusual artifice;" and Tu Yu amplifies this by saying: "In such a position, some scheme must be devised which will suit the circumstances, and if we can succeed in deluding the enemy, the peril may be escaped." This is exactly what happened on the famous occasion when Hannibal was hemmed in among the mountains on the road to Casilinum, and to all appearances entrapped by the dictator Fabius. The stratagem which Hannibal devised to baffle his foes was remarkably like that which T'ien Tan had also employed with success exactly 62 years before. When night came on, bundles of twigs were fastened to the horns of some 2,000 oxen and set on fire, the terrified animals being then quickly driven along the mountain side towards the passes which were beset by the enemy. The strange spectacle of these rapidly moving lights so alarmed and discomfited the Romans that they withdrew from their position, and Hannibal's army passed safely through the defile.

On desperate ground, fight.

For, as Chia Lin remarks: "if you fight with all your might, there is a chance of life; whereas death is certain if you cling to your corner."

15. Those who were called skillful leaders of old knew how to drive a wedge between the enemy's front and rear;

   More literally, "cause the front and rear to lose touch with each other."

   to prevent co-operation between his large and small divisions; to hinder the good troops from rescuing the bad, the officers from rallying their men.

16. When the enemy's men were scattered, they prevented them from concentrating, even when their forces were united, they managed to keep them in disorder.

17. When it was to their advantage, they made a forward move; when otherwise, they stopped still.

   Mei Yao-ch'ên connects this with the foregoing: "Having succeeded in thus dislocating the enemy, they would push forward in order to secure any advantage to be gained; if there was no advantage to be gained, they would remain where they were."

18. If asked how to cope with a great host of the enemy in orderly array and on the point of marching to the attack, I should say: "Begin by seizing something which your opponent holds dear; then he will be amenable to your will."

   Opinions differ as to what Sun Tzŭ had in mind. Ts'ao Kung thinks it is "some strategical advantage on which the enemy is depending." Tu Mu says: "The three things which an enemy is anxious to do, and on the accomplishment of which his success depends, are: (1) to capture our favorable positions; (2) to ravage our cultivated land; (3) to guard his own communications." Our object then must be to thwart his plans in these three directions and thus render him helpless. (*Cf.* ch. III, sect. 3.) . . . I agree with Ch'ên Hao, who says that [what should be seized] is any person or thing that may be of importance to the enemy. By boldly seizing the initiative in this way, you at once throw the other side on the defensive.

19. **Rapidity is the essence of war:**

According to Tu Mu, "this is a summary of leading principles in warfare," and he adds: "These are the profoundest truths of military science, and the chief business of the general." The following anecdotes, told by Ho Shih, show the importance attached to speed by two of China's greatest generals. In AD 227, Mêng Ta, governor of Hsin-ch'êng under the Wei Emperor Wên Ti, was meditating defection to the House of Shu, and had entered into correspondence with Chu-ko Liang, Prime Minister of that State. The Wei general Ssǔ-ma I was then military governor of Wan, and getting wind of Mêng Ta's treachery, he at once set off with an army to anticipate his revolt, having previously cajoled him by a specious message of friendly import. Ssǔ-ma's officers came to him and said: "If Mêng Ta has leagued himself with Wu and Shu, the matter should be thoroughly investigated before we make a move." Ssǔ-ma I replied: "Mêng Ta is an unprincipled man, and we ought to go and punish him at once, while he is still wavering and before he has thrown off the mask." Then, by a series of forced marches, be brought his army under the walls of Hsin-ch'êng with in a space of eight days. Now Mêng Ta had previously said in a letter to Chu-ko Liang: "Wan is 1,200 *li* from here. When the news of my revolt reaches Ssǔ-ma I, he will at once inform his imperial master, but it will be a whole month before any steps can be taken, and by that time my city will be well fortified. Besides, Ssǔ-ma I is sure not to come himself, and the generals that will be sent against us are not worth troubling about." The next letter, however, was filled with consternation: "Though only eight days have passed since I threw off my allegiance, an army is already at the city-gates. What miraculous rapidity is this!" A fortnight later, Hsin-ch'êng had fallen and Mêng Ta had lost his head. . . . In AD 621, Li Ching was sent from K'uei-chou in Ssǔ-Ch'üan to reduce the successful rebel Hsiao Hsien, who had set up as Emperor at the modern Ching-chou Fu in Hupeh. It was autumn, and the Yangtsze being then in flood, Hsiao Hsien never dreamt that his adversary would venture to come down through the gorges, and consequently made no preparations. But Li Ching embarked his army without loss of time, and was just about to start when the other generals implored him to postpone his departure until the river was in a less dangerous state for navigation. Li Ching replied: "To the soldier, overwhelming speed is of paramount importance, and he must never miss opportunities. Now is the time to strike, before Hsiao Hsien even knows that we have got an army together. If we seize the present moment when the river is in flood, we shall appear before his capital with startling suddenness, like the thunder which is heard before you have time to stop your ears against it. (See ch. VII, sect. 19.) This is the great principle in war. Even if he gets to know of our approach, he will have to levy his soldiers in such a hurry that they will not be fit to oppose us. Thus the full fruits of victory will be ours." All came about as he predicted, and Hsiao Hsien was obliged to surrender, nobly stipulating that his people should be spared and he alone suffer the penalty of death.

兵
法

take advantage of the enemy's unreadiness, make your way by unexpected routes, and attack unguarded spots.

20. **The following are the principles to be observed by an invading force: The further you penetrate into a country, the greater will be the solidarity of your troops, and thus the defenders will not prevail against you.**

21. **Make forays in fertile country in order to supply your army with food.**

    Cf. supra, sect. 13. Li Ch'üan does not venture on a note here.

22. Carefully study the well-being of your men,

[Well-being], according to Wang Hsi, means, "Pet them, humor them, give them plenty of food and drink, and look after them generally."

and do not overtax them. Concentrate your energy and hoard your strength.

. . . Ch'ên Hao recalls the line of action adopted in 224 BC by the famous general Wang Chien, whose military genius largely contributed to the success of the First Emperor. He had invaded the Ch'u State, where a universal levy was made to oppose him. But, being doubtful of the temper of his troops, he declined all invitations to fight and remained strictly on the defensive. In vain did the Ch'u general try to force a battle: day after day Wang Chien kept inside his walls and would not come out, but devoted his whole time and energy to winning the affection and confidence of his men. He took care that they should be well fed, sharing his own meals with them, provided facilities for bathing, and employed every method of judicious indulgence to weld them into a loyal and homogenous body. After some time had elapsed, he told off certain persons to find out how the men were amusing themselves. The answer was, that they were contending with one another in putting the weight and long-jumping. When Wang Chien heard that they were engaged in these athletic pursuits, he knew that their spirits had been strung up to the required pitch and that they were now ready for fighting. By this time the Ch'u army, after repeating their challenge again and again, had marched away eastwards in disgust. The Ch'in general immediately broke up his camp and followed them, and in the battle that ensued they were routed with great slaughter. Shortly afterwards, the whole of Ch'u was conquered by Ch'in, and the king Fu-Ch'u led into captivity.

Keep your army continually on the move,

In order that the enemy may never know exactly where you are. It has struck me, however, that the true reading might be "link your army together."

and devise unfathomable plans.

23. **Throw your soldiers into positions whence there is no escape, and they will prefer death to flight. If they will face death, there is nothing they may not achieve.**

    Chang Yü . . . quotes his favorite Wei Liao Tzŭ: "If one man were to run amok with a sword in the market-place, and everybody else tried to get out of his way, I should not allow that this man alone had courage and that all the rest were contemptible cowards. The truth is, that a desperado and a man who sets some value on his life do not meet on even terms."

    Officers and men alike will put forth their uttermost strength.

    Chang Yü says: "If they are in an awkward place together, they will surely exert their united strength to get out of it."

24. **Soldiers when in desperate straits lose the sense of fear. If there is no place of refuge, they will stand firm. If they are in the heart of a hostile country, they will show a stubborn front. If there is no help for it, they will fight hard.**

25. **Thus, without waiting to be marshaled, the soldiers will be constantly on the *qui vive*; without waiting to be asked, they will do your will;**

    Literally, "without asking, you will get."

兵
法

without restrictions, they will be faithful; without giving orders, they can be trusted.

26. **Prohibit the taking of omens, and do away with superstitious doubts. Then, until death itself comes, no calamity need be feared.**

The superstitious, "bound in to saucy doubts and fears," degenerate into cowards and "die many times before their deaths." Tu Mu quotes Huang Shih-kung: "'Spells and incantations should be strictly forbidden, and no officer allowed to inquire by divination into the fortunes of an army, for fear the soldiers' minds should be seriously perturbed.' The meaning is," he continues, "that if all doubts and scruples are discarded, your men will never falter in their resolution until they die."

27. **If our soldiers are not overburdened with money, it is not because they have a distaste for riches; if their lives are not unduly long, it is not because they are disinclined to longevity.**

Chang Yü has the best note on this passage: "Wealth and long life are things for which all men have a natural inclination. Hence, if they burn or fling away valuables, and sacrifice their own lives, it is not that they dislike them, but simply that they have no choice." Sun Tzŭ is slyly insinuating that, as soldiers are but human, it is for the general to see that temptations to shirk fighting and grow rich are not thrown in their way.

28. **On the day they are ordered out to battle, your soldiers may weep,**

The word in the Chinese is "snivel." This is taken to indicate more genuine grief than tears alone.

those sitting up bedewing their garments, and those lying down letting the tears run down their cheeks.

Not because they are afraid, but because, as Ts'ao Kung says, "all have embraced the firm resolution to do or die." We may remember that the heroes of the Iliad were equally childlike in showing their emotion. Chang Yü alludes to the mournful parting at the I River between Ching K'o and his friends, when the former was sent to attempt the life of the King of Ch'in (afterwards First Emperor) in 227 BC. The tears of all flowed down like rain as he bade them farewell and uttered the following lines: "The shrill blast is blowing, Chilly the burn; Your champion is going—Not to return."

But let them once be brought to bay, and they will display the courage of a Chu or a Kuei.

[Chu] was the personal name of Chuan Chu, a native of the Wu State and contemporary with Sun Tzŭ himself, who was employed by Kung-tzuŭ Kuang, better known as Ho Lu Wang, to assassinate his sovereign Wang Liao with a dagger which he secreted in the belly of a fish served up at a banquet. He succeeded in his attempt, but was immediately hacked to pieces by the king's bodyguard. This was in 515 BC. The other hero referred to, Ts'ao Kuei (or Ts'ao Mo), performed the exploit which has made his name famous 166 years earlier, in 681 BC. Lu had been thrice defeated by Ch'i, and was just about to conclude a treaty surrendering a large slice of territory, when Ts'ao Kuei suddenly seized Huan Kung, the Duke of Ch'i, as he stood on the altar steps and held a dagger against his chest. None of the duke's retainers dared to move a muscle, and Ts'ao Kuei proceeded to demand full restitution, declaring that Lu was being unjustly treated because she was a smaller and a weaker state. Huan Kung, in peril of his life, was obliged to consent, whereupon Ts'ao Kuei flung away his dagger and quietly resumed his place amid the terrified assemblage without having so much as changed color. As was to be expected, the Duke wanted afterwards to repudiate the bargain, but his wise old counselor Kuan Chung pointed out to him the impolicy of breaking his word, and the upshot was that this bold stroke regained for Lu the whole of what she had lost in three pitched battles.

29. The skillful tactician may be likened to the *shuai-jan*. Now the *shuai-jan* is a snake that is found in the Ch'ang mountains.

[Shuai-jan] means "suddenly" or "rapidly," and the snake in question was doubtless so called owing to the rapidity of its movements. Through this passage, the term in Chinese has now come to be used in the sense of "military maneuvers."

Strike at its head, and you will be attacked by its tail; strike at its tail, and you will be attacked by its head; strike at its middle, and you will be attacked by head and tail both.

30. Asked if an army can be made to imitate the *shuai-jan*,

That is, as Mei Yao-ch'ên says, "Is it possible to make the front and rear of an army each swiftly responsive to attack on the other, just as though they were part of a single living body?"

I should answer, Yes. For the men of Wu and the men of Yüeh are enemies;

Cf. ch. VI, sect. 21.

yet if they are crossing a river in the same boat and are caught by a storm, they will come to each other's assistance just as the left hand helps the right.

The meaning is: If two enemies will help each other in a time of common peril, how much more should two parts of the same army, bound together as they are by every tie of interest and fellow-feeling. Yet it is notorious that many a campaign has been ruined through lack of cooperation, especially in the case of allied armies.

兵
法

31. **Hence it is not enough to put one's trust in the tethering of horses, and the burying of chariot wheels in the ground.**

These quaint devices to prevent one's army from running away recall the Athenian hero Sophanes, who carried the anchor with him at the battle of Plataea, by means of which he fastened himself firmly to one spot. . . . It is not enough, says Sun Tzŭ, to render flight impossible by such mechanical means. You will not succeed unless your men have tenacity and unity of purpose, and, above all, a spirit of sympathetic cooperation. This is the lesson which can be learned from the *shuai-jan.*

32. **The principle on which to manage an army is to set up one standard of courage which all must reach.**

Literally, "level the courage (of all) as though (it were that of) one." If the ideal army is to form a single organic whole, then it follows that the resolution and spirit of its component parts must be of the same quality, or at any rate must not fall below a certain standard. Wellington's seemingly ungrateful description of his army at Waterloo as "the worst he had ever commanded" meant no more than that it was deficient in this important particular—unity of spirit and courage. Had he not foreseen the Belgian defections and carefully kept those troops in the background, he would almost certainly have lost the day.

33. **How to make the best of both strong and weak—that is a question involving the proper use of ground.**

Mei Yao-ch'ên's paraphrase is: "The way to eliminate the differences of strong and weak and to make both serviceable is to utilize accidental features of the ground." Less reliable troops, if posted in strong positions, will hold out as long as better troops on more exposed terrain. The advantage of position neutralizes the inferiority in stamina and courage. Col. Henderson, [in *The Science of War,*] says: "With all respect to the text books, and to the ordinary tactical teaching, I am inclined to think that the study of ground is often overlooked, and that by no means sufficient importance is attached to the selection of positions . . . and to the immense advantages that are to be derived, whether you are defending or attacking, from the proper utilization of natural features."

34. Thus the skillful general conducts his army just as though he were leading a single man, willy-nilly, by the hand.

Tu Mu says: "The simile has reference to the ease with which he does it."

35. It is the business of a general to be quiet and thus ensure secrecy; upright and just, and thus maintain order.

36. He must be able to mystify his officers and men by false reports and appearances,

Literally, "to deceive their eyes and ears."

and thus keep them in total ignorance.

Ts'ao Kung gives us one of his excellent apophthegms: "The troops must not be allowed to share your schemes in the beginning; they may only rejoice with you over their happy outcome." "To mystify, mislead, and surprise the enemy," is one of the first principles in war, as had been frequently pointed out. But how about the other process—the mystification of one's own men? Those who may think that Sun Tzǔ is over-emphatic on this point would do well to read Col. Henderson's remarks on Stonewall Jackson's Valley campaign: "The infinite pains," he says, "with which Jackson sought to conceal, even from his most trusted staff officers, his movements, his intentions, and his thoughts, a commander less thorough would have pronounced useless"—etc., etc. In the year AD 88 . . . "Pan Ch'ao took the field with 25,000 men from Khotan and other Central Asian states with the object of crushing Yarkand. The King of Kutcha replied by dispatching his chief commander to succor the place with an army drawn from the

kingdoms of Wen-su, Ku-mo, and Wei-t'ou, totaling 50,000 men. Pan Ch'ao summoned his officers and also the King of Khotan to a council of war, and said: 'Our forces are now outnumbered and unable to make head against the enemy. The best plan, then, is for us to separate and disperse, each in a different direction. The King of Khotan will march away by the easterly route, and I will then return myself towards the west. Let us wait until the evening drum has sounded and then start.' Pan Ch'ao now secretly released the prisoners whom he had taken alive, and the King of Kutcha was thus informed of his plans. Much elated by the news, the latter set off at once at the head of 10,000 horsemen to bar Pan Ch'ao's retreat in the west, while the King of Wen-su rode eastward with 8,000 horse in order to intercept the King of Khotan. As soon as Pan Ch'ao knew that the two chieftains had gone, he called his divisions together, got them well in hand, and at cock-crow hurled them against the army of Yarkand, as it lay encamped. The barbarians, panic-stricken, fled in confusion, and were closely pursued by Pan Ch'ao. Over 5,000 heads were brought back as trophies, besides immense spoils in the shape of horses and cattle and valuables of every description. Yarkand then capitulating, Kutcha and the other kingdoms drew off their respective forces. From that time forward, Pan Ch'ao's prestige completely overawed the countries of the west." In this case, we see that the Chinese general not only kept his own officers in ignorance of his real plans, but actually took the bold step of dividing his army in order to deceive the enemy.

37. **By altering his arrangements and changing his plans,**

Wang Hsi thinks that this means not using the same stratagem twice.

he keeps the enemy without definite knowledge.

Chang Yü, [quoting from another source] says: "The axiom, that war is based on deception, does not apply only to deception of the enemy. You must deceive even your own soldiers. Make them follow you, but without letting them know why."

By shifting his camp and taking circuitous routes, he prevents the enemy from anticipating his purpose.

38. At the critical moment, the leader of an army acts like one who has climbed up a height and then kicks away the ladder behind him. He carries his men deep into hostile territory before he shows his hand.

[L]iterally, "releases the spring" (see ch. V, sect. 15), that is, takes some decisive step which makes it impossible for the army to return—like Hsiang Yu, who sank his ships after crossing a river. Ch'ên Hao, followed by Chia Lin, understands the words less well as "puts forth every artifice at his command."

39. He burns his boats and breaks his cooking-pots; like a shepherd driving a flock of sheep, he drives his men this way and that, and nothing knows whither he is going.

Tu Mu says: "The army is only cognizant of orders to advance or retreat; it is ignorant of the ulterior ends of attacking and conquering."

40. To muster his host and bring it into danger:—this may be termed the business of the general.

Sun Tzŭ means that after mobilization there should be no delay in aiming a blow at the enemy's heart. . . . Note how he returns again and again to this point. Among the warring states of ancient China, desertion was no doubt a much more present fear and serious evil than it is in the armies of today.

41. The different measures suited to the nine varieties of ground;

Chang Yü says: "One must not be hide-bound in interpreting the rules for the nine varieties of ground."

兵
法

the expediency of aggressive or defensive tactics; and the fundamental laws of human nature: these are things that must most certainly be studied.

42. When invading hostile territory, the general principle is, that penetrating deeply brings cohesion; penetrating but a short way means dispersion.

Cf. supra, sect. 20.

43. When you leave your own country behind, and take your army across neighborhood territory, you find yourself on critical ground.

This "ground" is curiously mentioned in ch. VIII, sect. 2, but it does not figure among the Nine Situations or the Six Calamities in ch. X. One's first impulse would be to translate it "distant ground" . . . but this, if we can trust the commentators, is precisely what is not meant here. Mei Yao-ch'ên says it is "a position not far enough advanced to be called 'facile,' and not near enough to home to be 'dispersive,' but something between the two." . . . Wang Hsi says: "It is ground separated from home by an interjacent state, whose territory we have had to cross in order to reach it. Hence, it is incumbent on us to settle our business there quickly." He adds that this position is of rare occurrence, which is the reason why it is not included among the Nine Situations.

When there are means of communication on all four sides, the ground is one of intersecting highways.

44. When you penetrate deeply into a country, it is serious ground. When you penetrate but a little way, it is facile ground.

45. **When you have the enemy's strongholds on your rear, and narrow passes in front, it is hemmed-in ground. When there is no place of refuge at all, it is desperate ground.**

46. **Therefore, on dispersive ground, I would inspire my men with unity of purpose.**

    This end, according to Tu Mu, is best attained by remaining on the defensive, and avoiding battle. *Cf. supra*, sect. 11.

    On facile ground, I would see that there is close connection between all parts of my army.

    As Tu Mu says, the object is to guard against two possible contingencies: "(1) the desertion of our own troops; (2) a sudden attack on the part of the enemy." *Cf.* ch. VII, sect. 17. Mei Yao-ch'ên says: "On the march, the regiments should be in close touch; in an encampment, there should be continuity between the fortifications."

47. **On contentious ground, I would hurry up my rear.**

    This is Ts'ao Kung's interpretation. Chang Yü adopts it, saying: "We must quickly bring up our rear, so that head and tail may both reach the goal." That is, they must not be allowed to straggle up a long way apart. Mei Yao-ch'ên offers another equally plausible explanation: "Supposing the enemy has not yet reached the coveted position, and we are behind him, we should advance with all speed in order to dispute its possession." . . . *Cf.* ch. VIII, sect. 4. Ch'ên Hao, on the other hand, assuming that the enemy has had time to select his own ground, quotes ch. VI, sect. 1, where Sun Tzŭ warns us against coming exhausted to the attack. His own idea of the situation is rather vaguely expressed: "If there is a favorable position lying in front of you, detach a picked body of troops to occupy it, then if the enemy, relying on their numbers, come up to make a fight for it, *you may*

兵
法

fall quickly on their rear with your main body, and victory will be assured." It was thus, he adds, that Chao Shê beat the army of Ch'in.

48. **On open ground, I would keep a vigilant eye on my defenses. On ground of intersecting highways, I would consolidate my alliances.**

As Wang Hsi says, "fearing a surprise attack."

49. **On serious ground, I would try to ensure a continuous stream of supplies.**

The commentators take this as referring to forage and plunder, not, as one might expect, to an unbroken communication with a home base.

On difficult ground, I would keep pushing on along the road.

50. **On hemmed-in ground, I would block any way of retreat.**

"To make it seem that I meant to defend the position, whereas my real intention is to burst suddenly through the enemy's lines" (Mêng Shih); "in order to make my soldiers fight with desperation" (Mei Yao-ch'ên); "fearing lest my men be tempted to run away" (Wang Hsi). Tu Mu points out that this is the converse of ch. VII. sect. 36, where it is the enemy who is surrounded. In AD 532, Kao Huan, afterwards Emperor and canonized as Shên-wu, was surrounded by a great army under Êrh-chu Chao and others. His own force was comparatively small, consisting only of 2,000 horse and something under 30,000 foot. The lines of investment had not been drawn very closely together, gaps being left at certain points. But Kao Huan, instead of trying to escape, actually made a shift to block all the remaining outlets himself by driving into them a number of oxen and donkeys roped together. As soon as his officers and men saw that there was nothing for it but to conquer or die, their spirits rose to an extraordinary pitch of exaltation, and they charged with such desperate ferocity that the opposing ranks broke and crumbled under their onslaught.

On desperate ground, I would proclaim to my soldiers the hopelessness of saving their lives.

Tu Yu says: "Burn your baggage and impedimenta, throw away your stores and provisions, choke up the wells, destroy your cooking-stoves, and make it plain to your men that they cannot survive, but must fight to the death." Mei Yao-ch'ên says: "The only chance of life lies in giving up all hope of it." This concludes what Sun Tzǔ has to say about "grounds" and the "variations" corresponding to them. Reviewing the passages which bear on this important subject, we cannot fail to be struck by the desultory and unmethodical fashion in which it is treated. Sun Tzǔ begins abruptly in ch. VIII, sect. 2 to enumerate "variations" before touching on "grounds" at all, but only mentions five, namely nos. 7, 5, 8 and 9 of the subsequent list, and one that is not included in it. A few varieties of ground are dealt with in the earlier portion of ch. IX, and then ch. X sets forth six new grounds, with six variations of plan to match. None of these is mentioned again, though the first is hardly to be distinguished from ground no. 4 in the next chapter. At last, in ch. XI, we come to the Nine Grounds par excellence, immediately followed by the variations. This takes us down to sect. 14. In sects. 43–45, fresh definitions are provided for nos. 5, 6, 2, 8 and 9 (in the order given), as well as for the tenth ground noticed in ch. VIII; and finally, the nine variations are enumerated once more from beginning to end, all, with the exception of 5, 6 and 7, being different from those previously given. Though it is impossible to account for the present state of Sun Tzǔ's text, a few suggestive facts maybe brought into prominence: (1) Ch. VIII, according to the title, should deal with nine variations, whereas only five appear. (2) It is an abnormally short chapter. (3) Ch. XI is entitled The Nine Grounds. Several of these are defined twice over, besides which there are two distinct lists of the corresponding variations. (4) The length of the chapter is disproportionate, being double that of any other except ch. IX. I do not propose to draw any inferences from these facts, beyond the general conclusion

兵
法

that Sun Tzŭ's work cannot have come down to us in the shape in which it left his hands: ch. VIII is obviously defective and probably out of place, while XI seems to contain matter that has either been added by a later hand or ought to appear elsewhere.

51. **For it is the soldier's disposition to offer an obstinate resistance when surrounded, to fight hard when he cannot help himself, and to obey promptly when he has fallen into danger.**

[Chang Yü illuminates this passage by alluding] to the conduct of Pan Ch'ao's devoted followers in AD 73. The story runs thus in the *Hou Han Shu*: "When Pan Ch'ao arrived at Shan-shan, Kuang, the King of the country, received him at first with great politeness and respect; but shortly afterwards his behavior underwent a sudden change, and he became remiss and negligent. Pan Ch'ao spoke about this to the officers of his suite: 'Have you noticed,' he said, 'that Kuang's polite intentions are on the wane? This must signify that envoys have come from the Northern barbarians, and that consequently he is in a state of indecision, not knowing with which side to throw in his lot. That surely is the reason. The truly wise man, we are told, can perceive things before they have come to pass; how much more, then, those that are already manifest!' Thereupon he called one of the natives who had been assigned to his service, and set a trap for him, saying: 'Where are those envoys from the Hsiung-nu who arrived some day ago?' The man was so taken aback that between surprise and fear he presently blurted out the whole truth. Pan Ch'ao, keeping his informant carefully under lock and key, then summoned a general gathering of his officers, thirty-six in all, and began drinking with them. When the wine had mounted into their heads a little, he tried to rouse their spirit still further by addressing them thus: 'Gentlemen, here we are in the heart of an isolated region, anxious to achieve riches and honor by some great exploit. Now it happens that an ambassador from the Hsiung-nu arrived in this kingdom only a few days ago, and the result is that the respectful courtesy extended towards us by our royal

host has disappeared. Should this envoy prevail upon him to seize our party and hand us over to the Hsiung-nu, our bones will become food for the wolves of the desert. What are we to do?' With one accord, the officers replied: 'Standing as we do in peril of our lives, we will follow our commander through life and death.'"

52.   **We cannot enter into alliance with neighboring princes until we are acquainted with their designs. We are not fit to lead an army on the march unless we are familiar with the face of the country—its mountains and forests, its pitfalls and precipices, its marshes and swamps. We shall be unable to turn natural advantages to account unless we make use of local guides.**

These three sentences are repeated from ch. VII, sects. 12–14—in order to emphasize their importance, the commentators seem to think. I prefer to regard them as interpolated here in order to form an antecedent to the following words. With regard to local guides, Sun Tzŭ might have added that there is always the risk of going wrong, either through their treachery or some misunderstanding such as Livy records . . . Hannibal, we are told, ordered a guide to lead him into the neighborhood of Casinum, where there was an important pass to be occupied; but his Carthaginian accent, unsuited to the pronunciation of Latin names, caused the guide to understand Casilinum instead of Casinum, and turning from his proper route, he took the army in that direction, the mistake not being discovered until they had almost arrived.

53.   **To be ignored of any one of the following four or five principles does not befit a warlike prince.**

54.   **When a warlike prince attacks a powerful state, his generalship shows itself in preventing the concentration of the enemy's forces. He overawes his opponents, and their allies are prevented from joining against him.**

Mei Tao-ch'ên constructs one of the chains of reasoning that are so much affected by the Chinese: "In attacking a powerful state, if you can divide her forces, you will have a superiority in strength; if you have a superiority in strength, you will overawe the enemy; if you overawe the enemy, the neighboring states will be frightened; and if the neighboring states are frightened, the enemy's allies will be prevented from joining her." The following gives a stronger meaning [to this passage]: "If the great state has once been defeated (before she has had time to summon her allies), then the lesser states will hold aloof and refrain from massing their forces." Ch'ên Hao and Chang Yü take the sentence in quite another way. The former says: "Powerful though a prince may be, if he attacks a large state, he will be unable to raise enough troops, and must rely to some extent on external aid; if he dispenses with this, and with overweening confidence in his own strength, simply tries to intimidate the enemy, he will surely be defeated." Chang Yü puts his view thus: "If we recklessly attack a large state, our own people will be discontented and hang back. But if (as will then be the case) our display of military force is inferior by half to that of the enemy, the other chieftains will take fright and refuse to join us."

55. Hence he does not strive to ally himself with all and sundry, nor does he foster the power of other states. He carries out his own secret designs, keeping his antagonists in awe.

The train of thought appears to be this: Secure against a combination of his enemies, "he can afford to reject entangling alliances and simply pursue his own secret designs, his prestige enabling him to dispense with external friendships" (Li Ch'üan).

Thus he is able to capture their cities and overthrow their kingdoms.

This paragraph, though written many years before the Ch'in State became a serious menace, is not a bad summary of the policy by which the famous Six Chancellors gradually paved the way for her final triumph under Shih Huang Ti. Chang Yü, following up his previous note, thinks that Sun Tzŭ is condemning this attitude of cold-blooded selfishness and haughty isolation.

56. **Bestow rewards without regard to rule,**

Wu Tzŭ less wisely says: "Let advance be richly rewarded and retreat be heavily punished."

issue orders

Literally, "hang" or "post up."

without regard to previous arrangements;

"In order to prevent treachery," says Wang Hsi. The general meaning is made clear by Ts'ao Kung's quotation from the *Ssŭ-ma Fa*: "Give instructions only on sighting the enemy; give rewards when you see deserving deeds." Ts'ao Kung's paraphrase . . . I take to mean: "The final instructions you give to your army should not correspond with those that have been previously posted up." Chang Yü simplifies this into "your arrangements should not be divulged beforehand." And Chia Lin says: "there should be no fixity in your rules and arrangements." Not only is there danger in letting your plans be known, but war often necessitates the entire reversal of them at the last moment.

and you will be able to handle a whole army as though you had to do with but a single man.

Cf. supra, sect. 34.

57. **Confront your soldiers with the deed itself; never let them know your design.**

Literally, "do not tell them words"; i.e. do not give your reasons for any order. Lord Mansfield once told a junior colleague to "give no reasons" for his decisions, and the maxim is even more applicable to a general than to a judge.

When the outlook is bright, bring it before their eyes; but tell them nothing when the situation is gloomy.

58. **Place your army in deadly peril, and it will survive; plunge it into desperate straits, and it will come off in safety.**

These words of Sun Tzŭ were once quoted by Han Hsin in explanation of the tactics he employed in one of his most brilliant battles, already alluded to [in ch. IV, sect. 8]. In 204 BC, he was sent against the army of Chao, and halted ten miles from the mouth of the Ching-hsing pass, where the enemy had mustered in full force. Here, at midnight, he detached a body of 2,000 light cavalry, every man of which was furnished with a red flag. Their instructions were to make their way through narrow defiles and keep a secret watch on the enemy. "When the men of Chao see me in full flight," Han Hsin said, "they will abandon their fortifications and give chase. This must be the sign for you to rush in, pluck down the Chao standards and set up the red banners of Han in their stead." Turning then to his other officers, he remarked: "Our adversary holds a strong position, and is not likely to come out and attack us until he sees the standard and drums of the commander-in-chief, for fear I should turn back and escape through the mountains." So saying, he first of all sent out a division consisting of 10,000 men, and ordered them to form in line of battle with their backs to the River Ti. Seeing this maneuver, the whole army of Chao broke into loud laughter. By this time it was broad daylight, and Han Hsin, displaying the generalissimo's flag, marched out of the pass with drums beating, and was immediately engaged by the enemy. A great battle followed, lasting for some time; until at length Han Hsin and his colleague Chang Ni, leaving drums and banner on the field,

兵法

fled to the division on the river bank, where another fierce battle was raging. The enemy rushed out to pursue them and to secure the trophies, thus denuding their ramparts of men; but the two generals succeeded in joining the other army, which was fighting with the utmost desperation. The time had now come for the 2,000 horsemen to play their part. As soon as they saw the men of Chao following up their advantage, they galloped behind the deserted walls, tore up the enemy's flags and replaced them by those of Han. When the Chao army looked back from the pursuit, the sight of these red flags struck them with terror. Convinced that the Hans had got in and overpowered their king, they broke up in wild disorder, every effort of their leader to stay the panic being in vain. Then the Han army fell on them from both sides and completed the rout, killing a number and capturing the rest, amongst whom was King Ya himself. . . . After the battle, some of Han Hsin's officers came to him and said: "In the *Art of War* we are told to have a hill or tumulus on the right rear, and a river or marsh on the left front. (This appears to be a blend of Sun Tzŭ and T'ai Kung. See ch. IX, sect. 9, and note.) You, on the contrary, ordered us to draw up our troops with the river at our back. Under these conditions, how did you manage to gain the victory?" The general replied: "I fear you gentlemen have not studied the Art of War with sufficient care. Is it not written there: 'Plunge your army into desperate straits and it will come off in safety; place it in deadly peril and it will survive?' Had I taken the usual course, I should never have been able to bring my colleague round. What says the Military Classic—'Swoop down on the market-place and drive the men off to fight.' (This passage does not occur in the present text of Sun Tzŭ.) If I had not placed my troops in a position where they were obliged to fight for their lives, but had allowed each man to follow his own discretion, there would have been a general *débandade*, and it would have been impossible to do anything with them." The officers admitted the force of his argument, and said: "These are higher tactics than we should have been capable of."

59. **For it is precisely when a force has fallen into harm's way that is capable of striking a blow for victory.**

Danger has a bracing effect.

60. **Success in warfare is gained by carefully accommodating ourselves to the enemy's purpose.**

Ts'ao Kung says: "Feign stupidity"—by an appearance of yielding and falling in with the enemy's wishes. Chang Yü's note makes the meaning clear: "If the enemy shows an inclination to advance, lure him on to do so; if he is anxious to retreat, delay on purpose that he may carry out his intention." The object is to make him remiss and contemptuous before we deliver our attack.

61. **By persistently hanging on the enemy's flank,**

I understand the first four words to mean "accompanying the enemy in one direction." Ts'ao Kung says: "unite the soldiers and make for the enemy." But such a violent displacement of characters is quite indefensible.

兵
法

we shall succeed in the long run

Literally, "after a thousand *li*."

in killing the commander-in-chief.

Always a great point with the Chinese.

62. **This is called ability to accomplish a thing by sheer cunning.**

63. **On the day that you take up your command, block the frontier passes, destroy the official tallies,**

[These] were tablets of bamboo or wood, one half of which was issued as a permit or passport by the official in charge of a gate. . . .

When this half was returned to him, within a fixed period, he was authorized to open the gate and let the traveler through.

and stop the passage of all emissaries.

Either to or from the enemy's country.

64. **Be stern in the council-chamber,**

Show no weakness, and insist on your plans being ratified by the sovereign.

so that you may control the situation.

Mei Yao-ch'ên understands the whole sentence to mean: Take the strictest precautions to ensure secrecy in your deliberations.

65. **If the enemy leaves a door open, you must rush in.**

66. **Forestall your opponent by seizing what he holds dear,**

Cf. supra, sect. 18.

and subtly contrive to time his arrival on the ground.

Ch'ên Hao's explanation . . . is clear enough: "If I manage to seize a favorable position, but the enemy does not appear on the scene, the advantage thus obtained cannot be turned to any practical account. He who intends therefore, to occupy a position of importance to the enemy, must begin by making an artful appointment, so to speak, with his antagonist, and cajole him into going there as well." Mei Yao-ch'ên explains that this "artful appointment" is to be made through the medium of the enemy's own spies, who will carry back just the amount of information that we choose to give them. Then, having cunningly disclosed our intentions, "we must manage, though starting after the enemy, to arrive before him (ch. VII, sect. 4). We must start after him in order to ensure his marching thither; we must arrive before him in order to capture the place without trouble." Taken thus, the present passage lends some support to Mei Yao-ch'ên's interpretation of sect. 47.

67. **Walk in the path defined by rule,**

Chia Lin says: "Victory is the only thing that matters, and this cannot be achieved by adhering to conventional canons." It is unfortunate that this variant rests on very slight authority, for the sense yielded is certainly much more satisfactory. Napoleon, as we know, according to the veterans of the old school whom he defeated, won his battles by violating every accepted canon of warfare.

and accommodate yourself to the enemy until you can fight a decisive battle.

Tu Mu says: "Conform to the enemy's tactics until a favorable opportunity offers; then come forth and engage in a battle that shall prove decisive."

兵
法

68. **At first, then, exhibit the coyness of a maiden, until the enemy gives you an opening; afterwards emulate the rapidity of a running hare, and it will be too late for the enemy to oppose you.**

As the hare is noted for its extreme timidity, the comparison hardly appears felicitous. But of course Sun Tzŭ was thinking only of its speed. The words have been taken to mean: You must flee from the enemy as quickly as an escaping hare; but this is rightly rejected by Tu Mu.

THE ATTACK BY FIRE

Rather more than half the chapter (sects. 1–13) is devoted to the subject of fire, after which the author branches off into other topics.

孫
子

1. **Sun Tzŭ said: There are five ways of attacking with fire. The first is to burn soldiers in their camp;**

So Tu Mu. Li Ch'üan says: "Set fire to the camp, and kill the soldiers" (when they try to escape from the flames). Pan Ch'ao, sent on a diplomatic mission to the King of Shan-shan (see ch. XI, sect. 51), found himself placed in extreme peril by the unexpected arrival of an envoy from the Hsiung-nu [the mortal enemies of the Chinese]. In consultation with his officers, he exclaimed: "'Never venture, never win! The only course open to us now is to make an assault by fire on the barbarians under cover of night, when they will not be able to discern our numbers. Profiting by their panic, we shall exterminate them completely; this will cool the King's courage and cover us with glory, besides ensuring the success of our mission.' The officers all replied that it would be necessary to discuss the matter first with the Intendant. Pan Ch'ao then fell into a passion: 'It is today,' he cried, 'that our fortunes must be decided! The Intendant is only a humdrum civilian, who on hearing of our project will certainly be afraid, and everything will be brought to light. An inglorious death is no worthy fate for valiant warriors.' All then agreed to do as he wished. Accordingly, as soon as night came on, he and his little band quickly made their way to the barbarian camp. A strong gale was blowing at the time. Pan Ch'ao ordered ten of the party to take drums and hide behind the enemy's barracks, it being arranged that when they saw flames shoot up, they should begin drumming and yelling with all their might. The rest of his men, armed with bows and crossbows, he posted in ambuscade at the gate of the camp. He then set fire to the place from the windward side, whereupon a deafening noise of drums and shouting arose on the front and rear of the Hsiung-nu, who rushed out pell-mell in frantic disorder. Pan Ch'ao slew three of them with his own hand, while his companions cut off the heads of the envoy and thirty of his suite. The remainder, more than a hundred in all, perished in the flames. On the following day, Pan Ch'ao, divining his thoughts, said with uplifted hand: 'Although you did not go with us last night, I should not think, Sir, of taking sole credit for our exploit.' This satisfied Kuo Hsun, and Pan Ch'ao, having sent for Kuang, King of Shan-shan, showed him the head of the barbarian envoy. The whole kingdom was seized with fear and trembling, which Pan Ch'ao took steps to allay by issuing a public proclamation. Then, taking the king's sons as hostage, he returned to make his report to Tou Ku."

兵
法

the second is to burn stores;

Tu Mu says: "Provisions, fuel and fodder." In order to subdue the rebellious population of Kiangnan, Kao Keng recommended Wên Ti of the Sui dynasty to make periodic raids and burn their stores of grain, a policy which in the long run proved entirely successful.

the third is to burn baggage trains;

An example given is the destruction of Yüan Shao's wagons and impedimenta by Ts'ao Ts'ao in AD 200.

the fourth is to burn arsenals and magazines;

Tu Mu says that the things contained in "arsenals" and "magazines" are the same. He specifies weapons and other implements, bullion and clothing. *Cf.* ch. VII, sect. 11.

the fifth is to hurl dropping fire amongst the enemy.

[T]he interpretation which I have adopted is that given by Tu Yu says in the *T'ung Tien*: . . . "To drop fire into the enemy's camp. The method by which this may be done is to set the tips of arrows alight by dipping them into a brazier, and then shoot them from powerful crossbows into the enemy's lines."

2. **In order to carry out an attack, we must have means available;**

   T'sao Kung thinks that "traitors in the enemy's camp" are referred to. But Ch'ên Hao is more likely to be right in saying: "We must have favorable circumstances in general, not merely traitors to help us." Chia Lin says: "We must avail ourselves of wind and dry weather."

   the material for raising fire should always be kept in readiness.

   Tu Mu suggests "dry vegetable matter, reeds, brushwood, straw, grease, oil, etc." Here we have the material cause. Chang Yü says: "vessels for hoarding fire, stuff for lighting fires."

3. **There is a proper season for making attacks with fire, and special days for starting a conflagration.**

   A fire must not be begun "recklessly" or "at haphazard."

4. **The proper season is when the weather is very dry; the special days are those when the moon is in the constellations of the Sieve, the Wall, the Wing or the Cross-bar;**

   These are, respectively, the 7th, 14th, 27th, and 28th of the Twenty-eight Stellar Mansions, corresponding roughly to Sagittarius, Pegasus, Crater, and Corvus.

   for these four are all days of rising wind.

London Published Jan.ry 1.st 1804 by W. Mil.

5. In attacking with fire, one should be prepared to meet five possible developments:

6. (1) When fire breaks out inside to enemy's camp, respond at once with an attack from without.

7. (2) If there is an outbreak of fire, but the enemy's soldiers remain quiet, bide your time and do not attack.

The prime object of attacking with fire is to throw the enemy into confusion. If this effect is not produced, it means that the enemy is ready to receive us. Hence the necessity for caution.

8. (3) When the force of the flames has reached its height, follow it up with an attack, if that is practicable; if not, stay where you are.

Ts'ao Kung says: "If you see a possible way, advance; but if you find the difficulties too great, retire."

9. (4) If it is possible to make an assault with fire from without, do not wait for it to break out within, but deliver your attack at a favorable moment.

Tu Mu says that the previous paragraphs had reference to the fire breaking out (either accidentally, we may suppose, or by the agency of incendiaries) inside the enemy's camp. "But," he continues, "if the enemy is settled in a waste place littered with quantities of grass, or if he has pitched his camp in a position which can be burnt out, we must carry our fire against him at any seasonable opportunity, and not await on in hopes of an outbreak occurring within, for fear our opponents should themselves burn up the surrounding vegetation, and thus render our own attempts fruitless." The famous Li Ling once baffled the leader of the Hsiung-nu in this way. The latter, taking advantage of a favorable wind, tried to set fire to the

Chinese general's camp, but found that every scrap of combustible vegetation in the neighborhood had already been burnt down. On the other hand, Po-Ts'ai, a general of the Yellow Turban rebels, was badly defeated in AD 184 through his neglect of this simple precaution. "At the head of a large army he was besieging Ch'ang-she, which was held by Huang-fu Sung. The garrison was very small, and a general feeling of nervousness pervaded the ranks; so Huang-fu Sung called his officers together and said: 'In war, there are various indirect methods of attack, and numbers do not count for everything. (The commentator here quotes Sun Tzŭ, ch. V, sects. 5, 6 and 10.) Now the rebels have pitched their camp in the midst of thick grass which will easily burn when the wind blows. If we set fire to it at night, they will be thrown into a panic, and we can make a sortie and attack them on all sides at once, thus emulating the achievement of T'ien Tan.' That same evening, a strong breeze sprang up; so Huang-fu Sung instructed his soldiers to bind reeds together into torches and mount guard on the city walls, after which he sent out a band of daring men, who stealthily made their way through the lines and started the fire with loud shouts and yells. Simultaneously, a glare of light shot up from the city walls, and Huang-fu Sung, sounding his drums, led a rapid charge, which threw the rebels into confusion and put them to headlong flight."

兵
法

10. **(5) When you start a fire, be to windward of it. Do not attack from the leeward.**

Chang Yü, following Tu Yu, says: "When you make a fire, the enemy will retreat away from it; if you oppose his retreat and attack him then, he will fight desperately, which will not conduce to your success." A rather more obvious explanation is given by Tu Mu: "If the wind is in the east, begin burning to the east of the enemy, and follow up the attack yourself from that side. If you start the fire on the east side, and then attack from the west, you will suffer in the same way as your enemy."

11. **A wind that rises in the daytime lasts long, but a night breeze soon falls.**

Cf. Lao Tzǔ's saying [in the *Tao Te Ching*]: "A violent wind does not last the space of a morning." Mei Yao-ch'ên and Wang Hsi say: "A day breeze dies down at nightfall, and a night breeze at daybreak. This is what happens as a general rule." The phenomenon observed may be correct enough, but how this sense is to be obtained is not apparent.

12. **In every army, the five developments connected with fire must be known, the movements of the stars calculated, and a watch kept for the proper days.**

[Tu Mu says]: "We must make calculations as to the paths of the stars, and watch for the days on which wind will rise, before making our attack with fire." Chang Yü seems to interpret the text differently: "We must not only know how to assail our opponents with fire, but also be on our guard against similar attacks from them."

兵法

13. **Hence those who use fire as an aid to the attack show intelligence; those who use water as an aid to the attack gain an accession of strength.**

14. **By means of water, an enemy may be intercepted, but not robbed of all his belongings.**

Ts'ao Kung's note is: "We can merely obstruct the enemy's road or divide his army, but not sweep away all his accumulated stores." Water can do useful service, but it lacks the terrible destructive power of fire. This is the reason, Chang Yü concludes, why the former is dismissed in a couple of sentences, whereas the attack by fire is discussed in detail. Wu Tzǔ . . . speaks thus of the two elements: "If an army is encamped on low-lying marshy ground, from which the water cannot run off, and where the rainfall is heavy, it may be submerged by a flood. If an army is encamped in wild marsh lands thickly overgrown with weeds and brambles, and visited by frequent gales, it may be exterminated by fire."

15. **Unhappy is the fate of one who tries to win his battles and succeed in his attacks without cultivating the spirit of enterprise; for the result is waste of time and general stagnation.**

This is one of the most perplexing passages in Sun Tzŭ. . . . Ts'ao Kung says: "Rewards for good service should not be deferred a single day." And Tu Mu: "If you do not take opportunity to advance and reward the deserving, your subordinates will not carry out your commands, and disaster will ensue." . . . For several reasons, however, and in spite of the formidable array of scholars on the other side, I prefer the interpretation suggested by Mei Yao-ch'ên alone, whose words I will quote: "Those who want to make sure of succeeding in their battles and assaults must seize the favorable moments when they come and not shrink on occasion from heroic measures: that is to say, they must resort to such means of attack as fire, water and the like. What they must not do, and what will prove fatal, is to sit still and simply hold to the advantages they have got."

16. **Hence the saying: The enlightened ruler lays his plans well ahead; the good general cultivates his resources.**

Tu Mu quotes the following from [the *San Lueh*] . . . "The warlike prince controls his soldiers by his authority, kits them together by good faith, and by rewards makes them serviceable. If faith decays, there will be disruption; if rewards are deficient, commands will not be respected."

17. **Move not unless you see an advantage; use not your troops unless there is something to be gained; fight not unless the position is critical.**

    Sun Tzŭ may at times appear to be over-cautious, but he never goes so far in that direction as the remarkable passage in the *Tao Te Ching*, . . . "I dare not take the initiative, but prefer to act on the defensive; I dare not advance an inch, but prefer to retreat a foot."

18. **No ruler should put troops into the field merely to gratify his own spleen; no general should fight a battle simply out of pique.**

19. **If it is to your advantage, make a forward move; if not, stay where you are.**

    This is repeated from ch. XI, sect. 17. Here I feel convinced that it is an interpolation, for it is evident that sect. 20 ought to follow immediately on sect. 18.

20. **Anger may in time change to gladness; vexation may be succeeded by content.**

21. **But a kingdom that has once been destroyed can never come again into being;**

    The Wu State was destined to be a melancholy example of this saying.

    nor can the dead ever be brought back to life.

22. **Hence the enlightened ruler is heedful, and the good general full of caution. This is the way to keep a country at peace and an army intact.**

兵
法

THE USE OF SPIES

1. Sun Tzǔ said: Raising a host of a hundred thousand men
   and marching them great distances entails heavy loss on the
   people and a drain on the resources of the State. The daily
   expenditure will amount to a thousand ounces of silver.

   Cf. ch. II, sects. 1, 13, 14.

**There will be commotion at home and abroad, and men will
drop down exhausted on the highways.**

Chang Yü has the note: "We may be reminded of the saying: 'On
serious ground, gather in plunder.' Why then should carriage and
transportation cause exhaustion on the highways?—The answer
is, that not victuals alone, but all sorts of munitions of war have to
be conveyed to the army. Besides, the injunction to 'forage on the
enemy' only means that when an army is deeply engaged in hostile
territory, scarcity of food must be provided against. Hence, without being solely dependent on the enemy for corn, we must forage
in order that there may be an uninterrupted flow of supplies. Then,
again, there are places like salt deserts where provisions being unob-
tainable, supplies from home cannot be dispensed with."

**As many as seven hundred thousand families will be impeded
in their labor.**

Mei Yao-ch'ên says: "Men will be lacking at the plow-tail." The allusion is to the system of dividing land into nine parts . . . each consisting of about 15 acres, the plot in the center being cultivated on behalf
of the State by the tenants of the other eight. It was here also, so Tu
Mu tells us, that their cottages were built and a well sunk, to be used
by all in common. (See ch. II, sect. 12.) . . . In time of war, one of the
families had to serve in the army, while the other seven contributed
to its support. Thus, by a levy of 100,000 men (reckoning one able-
bodied soldier to each family) the husbandry of 700,000 families
would be affected.

2. **Hostile armies may face each other for years, striving for the victory which is decided in a single day. This being so, to remain in ignorance of the enemy's condition simply because one grudges the outlay of a hundred ounces of silver in honors and emoluments,**

"For spies" is of course the meaning, though it would spoil the effect of this curiously elaborate exordium if spies were actually mentioned at this point.

is the height of inhumanity.

Sun Tzŭ's agreement is certainly ingenious. He begins by adverting to the frightful misery and vast expenditure of blood and treasure which war always brings in its train. Now, unless you are kept informed of the enemy's condition, and are ready to strike at the right moment, a war may drag on for years. The only way to get this information is to employ spies, and it is impossible to obtain trustworthy spies unless they are properly paid for their services. But it is surely false economy to grudge a comparatively trifling amount for this purpose, when every day that the war lasts eats up an incalculably greater sum. This grievous burden falls on the shoulders of the poor, and hence Sun Tzŭ concludes that to neglect the use of spies is nothing less than a crime against humanity.

3. **One who acts thus is no leader of men, no present help to his sovereign, no master of victory.**

This idea, that the true object of war is peace, has its root in the national temperament of the Chinese. Even so far back as 597 BC, these memorable words were uttered by Prince Chuang of the Ch'u State: "The [Chinese] character for 'prowess' is made up of [the characters for] 'to stay' and 'a spear' (cessation of hostilities). Military prowess is seen in the repression of cruelty, the calling in of weapons, the preservation of the appointment of Heaven, the firm establishment of merit, the

兵
法

bestowal of happiness on the people, putting harmony between the princes, the diffusion of wealth."

4. **Thus, what enables the wise sovereign and the good general to strike and conquer, and achieve things beyond the reach of ordinary men, is** *foreknowledge.*

   That is, knowledge of the enemy's dispositions, and what he means to do.

5. **Now this foreknowledge cannot be elicited from spirits; it cannot be obtained inductively from experience,**

   Tu Mu's note makes the meaning clear: . . . "(knowledge of the enemy) cannot be gained by reasoning from other analogous cases."

   nor by any deductive calculation.

   Li Ch'üan says: "Quantities like length, breadth, distance and magnitude, are susceptible of exact mathematical determination; human actions cannot be so calculated."

6. **Knowledge of the enemy's dispositions can only be obtained from other men.**

   Mei Yao-ch'ên has rather an interesting note: "Knowledge of the spirit-world is to be obtained by divination; information in natural science may be sought by inductive reasoning; the laws of the universe can be verified by mathematical calculation: but the dispositions of an enemy are ascertainable through spies and spies alone."

7. **Hence the use of spies, of whom there are five classes: (1) Local spies; (2) inward spies; (3) converted spies; (4) doomed spies; (5) surviving spies.**

8. **When these five kinds of spy are all at work, none can discover the secret system. This is called "divine manipulation of the threads." It is the sovereign's most precious faculty.**

[According to Baden-Powell, in *Aids to Scouting*]: "Cromwell, one of the greatest and most practical of all cavalry leaders, had officers styled 'scout masters,' whose business it was to collect all possible information regarding the enemy, through scouts and spies, etc., and much of his success in war was traceable to the previous knowledge of the enemy's moves thus gained."

9. **Having *local spies* means employing the services of the inhabitants of a district.**

Tu Mu says: "In the enemy's country, win people over by kind treatment, and use them as spies."

10. **Having *inward spies*, making use of officials of the enemy.**

[Inward spies] includes both civil and military officials. Tu Mu enumerates the following classes as likely to do good service in this respect: "Worthy men who have been degraded from office, criminals who have undergone punishment; also, favorite concubines who are greedy for gold, men who are aggrieved at being in subordinate positions, or who have been passed over in the distribution of posts, others who are anxious that their side should be defeated in order that they may have a chance of displaying their ability and talents, fickle turncoats who always want to have a foot in each boat. Officials of these several kinds," he continues, "should be secretly approached and bound to one's interests by means of rich presents. In this way you will be able to find out the state of affairs in the enemy's country, ascertain the plans that are being formed against you, and moreover disturb the harmony and create a breach between the sovereign and his ministers." The necessity for extreme caution, however, in

dealing with "inward spies," appears from an historical incident related by Ho Shih: "Lo Shang, Governor of I-chou, sent his general Wei Po to attack the rebel Li Hsiung of Shu in his stronghold at P'i. After each side had experienced a number of victories and defeats, Li Hsiung had recourse to the services of a certain P'o-t'ai, a native of Wu-tu. He began to have him whipped until the blood came, and then sent him off to Lo Shang, whom he was to delude by offering to cooperate with him from inside the city, and to give a fire signal at the right moment for making a general assault. Lo Shang, confiding in these promises, marched out all his best troops, and placed Wei Po and others at their head with orders to attack at P'o-t'ai's bidding. Meanwhile, Li Hsiung's general, Li Hsiang, had prepared an ambuscade on their line of march; and P'o-t'ai, having reared long scaling-ladders against the city walls, now lighted the beacon-fire. Wei Po's men raced up on seeing the signal and began climbing the ladders as fast as they could, while others were drawn up by ropes lowered from above. More than a hundred of Lo Shang's soldiers entered the city in this way, every one of whom was forthwith beheaded. Li Hsiung then charged with all his forces, both inside and outside the city, and routed the enemy completely."

11. **Having *converted spies*, getting hold of the enemy's spies and using them for our own purposes.**

By means of heavy bribes and liberal promises detaching them from the enemy's service, and inducing them to carry back false information as well as to spy in turn on their own countrymen. . . . Ho Shih notes three occasions on which converted spies were used with conspicuous success: (1) by T'ien Tan in his defense of Chi-mo; (2) by Chao Shê on his march to O-yü; and (3) by the wily Fan Chu in 260 BC, when Lien P'o was conducting a defensive campaign against Ch'in. The King of Chao strongly disapproved of Lien P'o's cautious and dilatory methods, which had been unable to avert a series of minor disasters, and therefore lent a ready ear to the reports of his spies, who had secretly gone over to the enemy and were already in Fan Chu's pay. They said: "The only thing which causes Ch'in anxiety is lest Chao Kua should be made general. Lien P'o they consider an easy opponent, who is sure to be vanquished in the long run." Now this Chao Kua was a son of the famous Chao Shê. From his boyhood, he had been wholly engrossed in the study of war and military matters, until at last he came to believe that there was no commander in the whole Empire who could stand against him. His father was much disquieted by this overweening conceit, and the flippancy with which he spoke of such a serious thing as war, and solemnly declared that if ever Kua was appointed general, he would bring ruin on the armies of Chao. This was the man who, in spite of earnest protests from his own mother and the veteran statesman Lin Hsiang-ju, was now sent to succeed Lien P'o. Needless to say, he proved no match for the redoubtable Po Ch'i and the great military power of Ch'in. He fell into a trap by which his army was divided into two and his communications cut; and after a desperate resistance lasting 46 days, during which the famished soldiers devoured one another, he was himself killed by an arrow, and his whole force, amounting, it is said, to 400,000 men, ruthlessly put to the sword.

12. **Having *doomed spies*, doing certain things openly for purposes of deception, and allowing our spies to know of them and report them to the enemy.**

Tu Yu gives the best exposition of the meaning: "We ostentatiously do things calculated to deceive our own spies, who must be led to believe that they have been unwittingly disclosed. Then, when these spies are captured in the enemy's lines, they will make an entirely false report, and the enemy will take measures accordingly, only to find that we do something quite different. The spies will thereupon be put to death." . . . As an example of doomed spies, Ho Shih mentions the prisoners released by Pan Ch'ao in his campaign against Yarkand. . . . He also refers to T'ang Chien, who in AD 630 was sent by T'ai Tsung to lull the Turkish Khan Chieh-li into fancied security, until Li Ching was able to deliver a crushing blow against him. Chang Yü says that the Turks revenged themselves by killing T'ang Chien, but this is a mistake, for we read in both the Old and the New T'ang History . . . that he escaped and lived on until 656. Li I-chi played a somewhat similar part in 203 BC, when sent by the King of Han to open peaceful negotiations with Ch'i. He has certainly more claim to be described a "doomed spy," for the king of Ch'i, being subsequently attacked without warning by Han Hsin, and infuriated by what he considered the treachery of Li I-chi, ordered the unfortunate envoy to be boiled alive.

13. *Surviving spies*, **finally, are those who bring back news from the enemy's camp.**

This is the ordinary class of spies, properly so called, forming a regular part of the army. Tu Mu says: "Your surviving spy must be a man of keen intellect, though in outward appearance a fool; of shabby exterior, but with a will of iron. He must be active, robust, endowed with physical strength and courage; thoroughly accustomed to all sorts of dirty work, able to endure hunger and cold, and to put up with shame and ignominy." Ho Shih tells the following story of Ta-hsi Wu of the Sui dynasty: "When he was governor of Eastern Ch'in, Shên-wu of Ch'i made a hostile movement upon Sha-Yüan. The Emperor T'ai Tsu (Kao Tsu?) sent Ta-hsi Wu to spy upon the enemy. He was accompanied by two other men. All three were on horseback and wore the enemy's uniform. When it was dark, they dismounted a few hundred feet away from the enemy's camp and stealthily crept up to listen, until they succeeded in catching the passwords used in the army. Then they got on their horses again and boldly passed through the camp under the guise of night-watchmen; and more than once, happening to come across a soldier who was committing some breach of discipline, they actually stopped to give the culprit a sound cudgeling! Thus they managed to return with the fullest possible

兵
法

information about the enemy's dispositions, and received warm commendation from the Emperor, who in consequence of their report was able to inflict a severe defeat on his adversary."

14. **Hence it is that with none in the whole army are more intimate relations to be maintained than with spies.**

Tu Mu and Mei Yao-ch'ên point out that the spy is privileged to enter even the general's private sleeping-tent.

None should be more liberally rewarded. In no other business should greater secrecy be preserved.

Tu Mu gives a graphic touch: all communication with spies should be carried "mouth-to-ear." . . . The following remarks on spies may be quoted from Turenne, who made perhaps larger use of them than any previous commander: "Spies are attached to those who give them most, he who pays them ill is never served. They should never be known to anybody; nor should they know one another. When they propose anything very material, secure their persons, or have in your possession their wives and children as hostages for their fidelity. Never communicate anything to them but what is absolutely necessary that they should know.

15. **Spies cannot be usefully employed without a certain intuitive sagacity.**

Mei Yao-ch'ên says: "In order to use them, one must know fact from falsehood, and be able to discriminate between honesty and double-dealing." [Wang Hsi, in a different interpretation, thinks more along the lines of "intuitive perception" and "practical intelligence."] Tu Mu strangely refers these attributes to the spies themselves: "Before using spies we must assure ourselves as to their integrity of character and the extent of their experience and skill." But he continues: "A brazen face and a crafty disposition are more dangerous than mountains or rivers; it takes a man of genius to penetrate such." So that we are left in some doubt as to his real opinion on the passage.

16. **They cannot be properly managed without benevolence and straightforwardness.**

Chang Yü says: . . . "When you have attracted them by substantial offers, you must treat them with absolute sincerity; then they will work for you with all their might."

17. **Without subtle ingenuity of mind, one cannot make certain of the truth of their reports.**

Mei Yao-ch'ên says: "Be on your guard against the possibility of spies going over to the service of the enemy."

18. **Be subtle! be subtle! and use your spies for every kind of business.**

Cf. ch. VI. sect. 9.

19. **If a secret piece of news is divulged by a spy before the time is ripe, he must be put to death together with the man to whom the secret was told.**

[W]ord for word, the translation here is: "If spy matters are heard before [our plans] are carried out," etc. Sun Tzŭ's main point in this passage is: Whereas you kill the spy himself "as a punishment for letting out the secret," the object of killing the other man is only, as Ch'ên Hao puts it, "to stop his mouth" and prevent news leaking any further. If it had already been repeated to others, this object would not be gained. Either way, Sun Tzŭ lays himself open to the charge of inhumanity, though Tu Mu tries to defend him by saying that the man deserves to be put to death, for the spy would certainly not have told the secret unless the other had been at pains to worm it out of him.

20. **Whether the object be to crush an army, to storm a city, or to assassinate an individual, it is always necessary to begin by finding out the names of the attendants, the aides-de-camp,**

[Interpreted by Tu Yu as] "those whose duty it is to keep the general supplied with information," which naturally necessitates frequent interviews with him.

the door-keepers and sentries of the general in command. Our spies must be commissioned to ascertain these.

As the first step, no doubt towards finding out if any of these important functionaries can be won over by bribery.

21. **The enemy's spies who have come to spy on us must be sought out, tempted with bribes, led away and comfortably housed. Thus they will become converted spies and available for our service.**

22. **It is through the information brought by the converted spy that we are able to acquire and employ local and inward spies.**

[Tu Yu says]: "through conversion of the enemy's spies we learn the enemy's condition." And Chang Yü says: "We must tempt the converted spy into our service, because it is he that knows which of the local inhabitants are greedy of gain, and which of the officials are open to corruption."

23. **It is owing to his information, again, that we can cause the doomed spy to carry false tidings to the enemy.**

"Because the converted spy knows how the enemy can best be deceived" (Chang Yü).

24. **Lastly, it is by his information that the surviving spy can be used on appointed occasions.**

25. **The end and aim of spying in all its five varieties is knowledge of the enemy; and this knowledge can only be derived, in the first instance, from the converted spy.**

As explained in sects. 22–24. He not only brings information himself, but makes it possible to use the other kinds of spy to advantage.

Hence it is essential that the converted spy be treated with the utmost liberality.

26. **Of old, the rise of the Yin dynasty**

Sun Tzŭ means the Shang dynasty, founded in 1766 BC. Its name was changed to Yin by P'an Kêng in 1401.

was due to Î Chih

Better known as Î Yin, the famous general and statesman who took part in Ch'êng T'ang's campaign against Chieh Kuei.

who had served under the Hsia. Likewise, the rise of the Chou dynasty was due to Lü Ya

Lü Shang . . . rose to high office under the tyrant Chou Hsin, whom he afterwards helped to overthrow. Popularly known as T'ai Kung, a title bestowed on him by Wen Wang, he is said to have composed a treatise on war, erroneously identified with the *Liu T'ao*.

who had served under the Yin.

There is less precision in the Chinese than I have thought it well to introduce into my translation, and the commentaries on the passage are by no means explicit. But, having regard to the context, we can hardly doubt that Sun Tzŭ is holding up Î Chih and Lü Ya as illustrious examples of the converted spy, or something closely analogous. His suggestion is, that the Hsia and Yin dynasties were upset owing

兵
法

to the intimate knowledge of their weaknesses and shortcomings which these former ministers were able to impart to the other side. Mei Yao-ch'ên appears to resent any such aspersion on these historic names: "Î Yin and Lü Ya," he says, "were not rebels against the Government. Hsia could not employ the former, hence Yin employed him. Yin could not employ the latter, hence Chou employed him. Their great achievements were all for the good of the people." Ho Shih is also indignant: "How should two divinely inspired men such as Î and Lü have acted as common spies? Sun Tzŭ's mention of them simply means that the proper use of the five classes of spies is a matter which requires men of the highest mental caliber like Î and Lü, whose wisdom and capacity qualified them for the task. The above words only emphasize this point." Ho Shih believes then that the two heroes are mentioned on account of their supposed skill in the use of spies. But this is very weak.

27. **Hence it is only the enlightened ruler and the wise general who will use the highest intelligence of the army for purposes of spying, and thereby they achieve great results.**

Tu Mu closes with a note of warning: "Just as water, which carries a boat from bank to bank, may also be the means of sinking it, so reliance on spies, while productive of great results, is oft-times the cause of utter destruction."

Spies are a most important element in water, because on them depends an army's ability to move.

Chia Lin says that an army without spies is like a man with ears or eyes.

THE LIFE AND TIMES OF SUN TZŬ

BCE

1046 The Zhou dynasty is founded in China. The Zhou kings rule through a feudal system, in which China is divided into hundreds of hereditary fiefdoms.

CA. 770–481 This is the time known as the Spring and Autumn Period, during which the vassal states of the Zhou gradually attain more power through the annexation of smaller states and reduce the Zhou king to a figurehead.

551 The birth of the Chinese philosopher Confucius.

CA. 544 The traditionally accepted time of the birth of Sun Tzŭ.

514–496 The reign of Ho Lu (Helü), the king of Wu, who according to several sources, tested Sun Tzŭ by giving him command of 180 ladies of the palace before making him a general.

505–473 Giles' estimation of the writing of *The Art of War*.

CA. 496 The traditionally accepted year of Sun Tzŭ's death.

479 The death of Confucius.

CA.481–221 This is the time in Zhou dynasty China known as the Warring States Period, during which seven states ruled by kings consolidate and centralize power. As a result, wars grow in scope.

CA. 380 The birth of the military strategist and alleged descendant of Sun Tzŭ, Sun Pin (Sun Bin), to whom many of the later stories about Sun Tzŭ possibly refer.

MID-4TH CENTURY Modern critical research estimates that the writing of *The Art of War* began around this time, possibly taking seven decades and involving several authors.

CA. 316 The death of Sun Pin.

230 The state of Qin begins a rapid conquest of the other six Warring states and unifies China.

221 The short-lived Qin dynasty begins, and the Warring States Period ends.

CA. 97 Ssŭ-ma Ch'ien (Sima Qian) writes a short biography of Sun Tzŭ in his *Shiji*, or the *Records of the Grand Historian*. Central to the biography is the likely apocryphal tale of the meeting between King Ho Lu and Sun Tzŭ, which leads to the execution of the king's favorite concubines.

CE

1ST–2ND CENTURY The tradition of commentaries on *The Art of War* begins to grow in China, eventually reaching over two hundred in number.

8TH CENTURY *The Art of War* is introduced in Japan. The work is claimed to have influenced Japanese generals and samurai alike, and aided in the unification of Japan.

1067–1085 The reign of Emperor Shen Tsung of the Sung dynasty. Shen Tsung canonizes *The Art of War* as a part of the "Seven Military Classics" of ancient China, which becomes mandatory reading for army and government entrance examinations.

11TH–12TH CENTURY The eleven most important commentaries are compiled into a single volume by Ji Tianbao.

1195–1224 The reign of the emperor Ning-tsung (Ningzong) of Sung. The first woodblock-printed editions of the "Seven Military Classics," including *The Art of War*, originate from the time of the Sung dynasty. The earliest extant copy is from Ning-tsung's reign.

CA. 1750 The first classical Japanese translation of *The Art of War*, *Sonshi kokujikai*, by Butsu Mokei, is published in Kyoto. Translations into modern Japanese will not appear until the twentieth century.

1772 The first French translation of *The Art of War*, *Les treize articles sur l'Art Militaire, ouvrage composé en Chinois par Sun-tse*, by the Jesuit Father Jean-Joseph-Marie Amiot, is published in Paris, making the work available in a Western language.

CA. **1800** The scholar Sun Hsing-yen (1752–1818) puts together the Chinese version of *The Art of War* that will be considered the standard edition until the 1970s. (Most modern English editions—including Giles'—are based on this one.)

1799–1815 The Napoleonic Wars, during which Napoleon Bonaparte conquers most of Western Europe, take place. There is a widespread myth that Napoleon carried *The Art of War* with him on his campaigns, but in reality, there is little historical proof of him ever having studied Sun Tzǔ.

1860 The first Russian translation of *The Art of War* is published.

1905 Captain Everard Ferguson Calthrop of the British Army translates *The Art of War* from Japanese. This first-ever English translation is published in Japan.

1910 Lionel Giles publishes his English translation, which severely criticizes the previous ones. This translation becomes the standard and most widespread English version.

Bruno Navarra publishes his German version, *Das Buch vom Kriege—der Militärklassiker der Chinesen*.

1939 Germany attacks Poland, marking the beginning of World War II. Many believe that the Nazi High Command read *The Art of War*, but there is little or no evidence of this being the case.

1941 The Japanese attack Pearl Harbor. The architect of the attack, Admiral Isoroku Yamamoto, not only attended Harvard University, but also studied Sun Tzǔ.

1948 Lieutenant-general of the Imperial Japanese Army, Muto Akira, is executed for war crimes committed during World War II. Before the war, Muto had written an influential comparative study of Sun Tzǔ and Clausewitz for the Imperial Staff College.

1949 Mao Tse-tung (Mao Zedong) establishes the People's Republic of China after defeating the Nationalists in the Chinese Civil War. *The Art of War* greatly influenced the Communist strategy of protracted warfare, which in turn inspired other revolutionaries worldwide.

兵
法

1963 *The Art of War* is translated into English by retired US Marine Corps brigadier general Samuel B. Griffith. This edition contains a foreword by the noted British strategic thinker Captain B. H. Liddell Hart, who emphasizes the similarities between Sun Tzŭ's ideas and his own strategy of the "indirect approach."

1972 An almost complete bamboo strip version of the thirteen chapters of *The Art of War* dating to the late second century BCE is found at Lin-i in Shantung Province in China, proving that the text existed in its current length for more than two thousand years.

1975 After two decades of fighting, the capital of South Vietnam falls to the communist forces, ending the Vietnam War. The architect of victory, General Vo Nguyen Giap, was an avid reader of Sun Tzŭ and Mao Tse-tung.

1980s Japan rises as an economic rival to the United States. Business comes to be seen, in some quarters, as an extension of war. It is no wonder that *The Art of War* begins to be used as a business manual, first in Japan, then in other countries as well.

1984 *The Art of War* is introduced into the US National War College curriculum as part of an 883-page collection of texts on strategy called *The Art and Practice of Military Strategy.*

1989 The US Marine Corps doctrine *Warfighting* states that "*The Art of War* should be on every Marine officer's list of essential reading. . . . *The Art of War* is every bit as valuable today as when it was written."

1991 The Persian Gulf War occurs. Coalition strategists, including the commander of the coalition forces, General Norman Schwarzkopf, are often said to have practiced the major principles propounded in *The Art of War* in this military campaign.

1993 J. H. Huang publishes a new translation of *The Art of War.* A later edition contains a foreword by Lieutenant General David W. Barno, a former Afghanistan commander.

Ralph D. Sawyer publishes *The Seven Military Classics of Ancient China,* which is the first English translation of the collection and includes a new translation of *The Art of War.*

EXAMPLES OF MODERN TRANSLATIONS

Modern English translations of Sun Tzŭ can be very different. The text allows translators significant room for interpretation and the intrusion of personal and professional preferences. The following contrasting passages from different modern translations provide a taste of these differences. Readers can probably recognize, with little difficulty, the translations by the Victorian scholar (Lionel Giles), the Communist general (Tao Hanzhang), the Marine Corps general (Samuel B. Griffith), the professor of Chinese literature (John Minford), the interpreter bringing the text up-to-date (J. H. Huang) and, perhaps also, the modern consultant (Ralph Sawyer).

The five constant factors, presented in chapter 1, that govern the art of war:

GILES	GRIFFITH	SAWYER
The Moral Law	moral influence	the Tao
Heaven	weather	Heaven
Earth	terrain	Earth
The Commander	command	generals
Method and discipline	doctrine	the laws [for military organization and discipline]

MINFORD	HUANG	HANZHANG
The Way	leadership	politics
Heaven	cyclic natural occurrences	weather
Earth	geographical factors	terrain
Command	commandership	commander
Discipline	rules	doctrine

One of the three ways in which a ruler can bring misfortune on his army (chapter 3):

GILES	GRIFFITH	SAWYER
"By attempting to govern an army in the same way as he administers a kingdom, being ignorant of the conditions which obtain in an army. This causes restlessness in the soldier's minds."	"When ignorant of military affairs, to participate in their administration. This causes officers to be perplexed."	"He does not understand the Three Armies' military affairs but [directs them] in the same way as his [civil] administration. The officers will become confused."

MINFORD	HUANG	HANZHANG
"Ignorant interference In military decisions Confuses Officers and men"	"By not knowing the armies' affairs yet interfering with the armies' policies; the armies' warriors will be leery."	"When ignorant of military affairs, to interfere in their administration. This causes the officers to be perplexed."

About leadership (chapter 4):

GILES	GRIFFITH	SAWYER
"The consummate leader cultivates the moral law, and strictly adheres to method and discipline; thus it is in his power to control success."	"Those skilled in war cultivate the *Tao* and preserve the laws and are therefore able to formulate victorious policies."	"One who excels at employing the military cultivates the Tao and preserves the laws; therefore, he is able to be the regulator of victory and defeat."

MINFORD	HUANG	HANZHANG
"The Skillful Strategist Cultivates The Way And preserves The law; Thus he is master Of victory and defeat."	"Those sophisticated at strategy cultivate leadership and sustain rules so that they can be the controllers of victory and of defeat."	"Those skilled in war cultivate politics, preserve the laws and institutions, and are therefore able to formulate victorious policies."

About guaranteeing victory (chapter 10):

GILES	GRIFFITH	HUANG
"Hence the saying: If you know the enemy and know yourself, your victory will not stand in doubt; if you know Heaven and know Earth, you may make your victory complete."	"And therefore I say: Know the enemy, know yourself; your victory will never be endangered. Know the ground, know the weather; your victory will then be total."	"By perceiving the enemy and perceiving ourselves, victory thereby has no unforeseen risks. By perceiving the geographical factors and perceiving the cyclic natural occurrences, victory thereby is complete."

About the role of the general (chapter 11):

GILES	HUANG	HANZHANG
"It is the business of a general to be quiet and thus ensure secrecy; upright and just, and thus maintain order. He must be able to mystify his officers and men by false reports and appearances, and thus keep them in total ignorance. By altering his arrangements and changing his plans, he keeps the enemy without definite knowledge."	"Concerns in commanding armies are: scheming in impenetrability and being impartial in supervision; having the capability to subjugate the eyes and ears of warriors and soldiers, making them suppress their senses; modifying their assignments and adapting their plans, making the people suppress their judgment."	"It is the business of the general to be serene and inscrutable, impartial and self-controlled. He should be capable of keeping his officers and men in ignorance of his plans. He changes his methods and alters his plans so that people have no knowledge of what he aims at."

兵
法

BASED ON THE BOOK

Interestingly, *The Art of War* has been adapted several times into the comic book format. Tsai Chih Chung's *Sunzi Speaks: The Art of War* (1994) is one example. Published in a series of comics depicting Asian thinking, it features the original Chinese script with illustrations and an English translation. Other comic versions include *Sun Zi's Tactics* by Lee Chi Ching, which won Japan's International Manga Award in 2007. Yet another version was published in 2011 by SmarterComics; it was advertised with the promise that readers could "learn the secrets of succeeding in any conflict without having to read the ancient text."

孫
子

The ancient Chinese work has also surfaced in several movies. In the 1987 Hollywood film *Wall Street* each of the two main characters, Bud Fox (played by Charlie Sheen) and Gordon Gekko (played by Michael Douglas, who won the Academy Award for Best Actor for the role), refers to Sun Tzŭ. The movie became iconic and increased the popularity of *The Art of War* in the business world, where it had already generated interest as the work was viewed as having contributed in part to the strong Japanese economy. Sun Tzŭ also appears in the James Bond movie *Die Another Day*, released in 2002. In the film, parts of *The Art of War* are recited a few times by the North Korean Colonel Tan-Sun Moon, who has undergone DNA replacement therapy and reappears as Sir Gustav Graves, played by Toby Stephens. The film also includes a copy of *The Art of War* with a knife embedded in it. (Note that the 2000 action film entitled *The Art of War*, starring Wesley Snipes, has no connection to Sun Tzŭ other than its title, nor do the movie's two sequels.)

The Art of War has also been referred to on the small screen. In a 2001 episode of the television series *The Sopranos*, mob boss Tony Soprano (played by James Gandolfini) hears from his therapist about *The Art of War.* He later tells her that he has not only read the book, but been greatly influenced by it. Comparing Sun Tzŭ and Machiavelli, he states that the former is much better in strategy than the latter. Reportedly, Oxford University Press immediately printed 25,000 copies of *The Art of War*, which soon became an international bestseller. This success even produced a new version of the book called *The Art of War: As Featured on "The Sopranos."*

In 2009 the History Channel released a special about Sun Tzŭ, whom promotional materials for the show referred to as "the Nostradamus of warfare." The program strove to show how the book influenced a number of great historical battles "from Rome to the Civil War, WWII, and present day." It did not take long for the DVD of the show to appear on the History Channel list of Best Sellers.

The influence of *The Art of War* has even spread to the music world. The title of the Manic Street Preachers' 2001 album *Know Your Enemy* derived its name from the ancient text. (The working title of the album had been *Fidel*, after the Cuban revolutionary leader.)

It will probably come as no surprise that *The Art of War* has been referenced in several books. In 1999, two colonels of China's People's Liberation Army published a book called *Unrestricted Warfare*, which was released in English with the ominous subtitle *China's Master Plan to Destroy America*. The book considers a variety of non-military strategies that can be used against an opponent that is stronger militarily. The authors credit Sun Tzŭ with being the first creator of the principles of war, which are "still articles of faith for modern strategists."

The 2009 book *Horse Soldiers* by Doug Stanton follows US Special Forces and CIA advisors in northern Afghanistan in the months following the September 11, 2001, terrorist attacks. One member of the advisory team is reading *The Art of War*, and passages from the book are used to illustrate the realities of the initial military campaign in

兵
法

Afghanistan. Al-Qaeda seems to have done different reading, as an Everyman edition of Clausewitz's *On War* was found in an Afghan safe house after the invasion.

Of course, *The Art of War* has had an impact on many strategists and military leaders. After September 11, 2001, the CIA agent Henry "Hank" Crumpton had an important role in the response of the United States to the attacks by al-Qaeda. Crumpton is allegedly the man behind the strategy that ousted the Taliban from Afghanistan by combining small US intelligence and military teams and air assets with Afghan Northern Alliance ground troops. He has declared that *The Art of War* was one of the main influences on his thinking.

孫
子

FURTHER READING

Ames, Roger, trans. and ed. Sun-Tzŭ, *The Art of Warfare: The First English Translation Incorporating the Recently Discovered Yin-ch'üeh-shan Texts*. New York: Ballantine, 1993.

Clausewitz, Carl von. *On War*. The Barnes & Noble Library of Essential Reading. New York: Barnes & Noble, Inc., 2004.

Giles, Lionel, trans. and ed. *Sun Tzŭ on the Art of War: The Oldest Military Treatise in the World*. London: Luzac, 1910.

Handel, Michael I. *Masters of War: Classical Strategic Thought*. 3rd ed. London: Frank Cass, 2001.

Hanzhang, Tao. *Sun Tzŭ's Art of War: The Modern Chinese Interpretation*. Translated by Yuan Shibing. New York: Sterling Publishing, 1987.

Lau, D. C. "Some Notes on the Sun-tzu." *Bulletin of the School of African and Oriental Studies* 28, no. 2 (1965): 318–335.

——— and Roger T. Ames, trans. and eds. *Sun Pin, The Art of Warfare*. New York: Ballantine, 1996.

Loewe, Michael, and Edward L. Shaughnessy, eds. *The Cambridge History of China: From the Origins of Civilization to 221 BC*. Cambridge: Cambridge University Press, 1999.

Mair, Victor H., trans. *The Art of War: Sun Zi's Military Methods*. New York: Columbia University Press, 2007.

Sawyer, Ralph D., with Mei-chün Sawyer, trans. and eds. *The Seven Military Classics of Ancient China*. Boulder, CO: Westview, 1993.

Sun Tzŭ. *The Art of War*. Translated by Samuel B. Griffith. Oxford: Oxford University Press, 1963.

———. *The Art of War*. Edited and translated by J. H. Huang. New York: HarperPerennial, 2008.

———. *The Art of War*. Edited and translated by John Minford. New York: Penguin, 2003.

———. *The Art of War*. Translated by Ralph D. Sawyer. Boulder, CO: Westview Press, 1994.

Yates, Robin. "New Light on Ancient Chinese Military Texts: Notes on Their Nature and Evolution and the Development of Military Specialization in Warring States China." *T'oung Pao 74*, nos. 4–5 (1988): 212–248.

兵
法

Image Credits

孫
子